KW-480-436

STEAM RENAISSANCE

The decline and rise
of Steam Locomotives
in Britain

Tom Heavyside

DAVID & CHARLES
Newton Abbot London North Pomfret (VT)

Dedication
**To my mother and father
for much help untold
over very many years.**

British Library Cataloguing in Publication Data

Heavyside, Tom
  Steam renaissance.
  1. Locomotives – Great Britain – History
  I. Title
  625.2'61'0941   TJ603.4.G7

  ISBN 0-7153-8294-2

Book designed by Michael Head

© G. T. Heavyside 1984
All rights reserved. No part of this
publication may be reproduced, stored
in a retrieval system, or transmitted,
in any form or by any means, electronic,
mechanical, photocopying, recording or
otherwise, without the prior permission
of David & Charles (Publishers) Limited

Phototypeset in Plantin
by Typesetters (Birmingham) Limited
Smethwick, Warley
and printed in Great Britain
by Butler & Tanner Ltd, Frome
for David & Charles (Publishers) Limited
Brunel House Newton Abbot Devon

Published in the United States of America
by David & Charles Inc
North Pomfret Vermont 05053 USA

Acknowledgements
I am mindful of the tremendous assistance I have
received from many people during the completion
of this book, either by the supply of information or
the provision of facilities for photography. The list
is far too long to mention everyone by name but my
special thanks are due to Edmund Brain, Murray
Brown, Derek Mercer, Christina Siviter, David
Williams, and Alastair Wood. Thanks are due also
to those who have helped plug the gaps in my own
photographic coverage, the prints used being
acknowledged individually; the remainder are the
work of the author. Once again I am indebted to
Marlene McPherson for her splendid work on the
typewriter, and the maps were drawn by Derek
Waugh.

TOM HEAVYSIDE
Bolton

CONTENTS

PROLOGUE

Think back about 20 years to the early 1960s when the steam era in Britain was fast drawing to a close. Some areas were already steamless, Kent and East Anglia for example. Electrification was creeping steadily south and would be at Euston in a year or two. For the enthusiast – or should it be the steam enthusiast – it was a decidedly depressing time as last of class specials marked the end of yet more types which went their ways to the scrapyards.

Yet there were still opportunities to sample the unusual, and cross country journeys could often produce some surprises. A Monday in May 1963 was just one of those days when one enthusiast's diary noted some remarkable contrasts during a journey from Westhoughton (on the Bolton–Wigan line) to Bristol via the North and West route. The first stage to Manchester Victoria was with steam behind Stanier 2-6-4T No 42589 on a local stopping train. In Manchester it was a quick trip across the city to the then recently modernised Piccadilly station to board a Cardiff train.

Manchester Piccadilly presented a complete contrast to the aptly named smoky Victoria, with diesel and electric power much in evidence; at the head of the train stood 25kV ac overhead electric locomotive No E3062 (now No 85007) which swiftly reached Crewe, where it was replaced by English Electric Type 4 diesel-electric No D308 (later No 40108) onwards to Shrewsbury. Here at this important railway crossroads as the diesel was being uncoupled, Stanier Jubilee class 4-6-0 No 45572 *Eire* with a full head of steam eagerly waited to take over.

In the front coach, with the windows fully open, the ear-splitting roar from *Eire* as it grappled with the long taxing climb of over 12 miles to Church Stretton, was enthralling to that enthusiast – if not perhaps to the mere travellers anxious to get from one place to another and oblivious to the aural differences between various classes of steam locomotive. Running was much freer along the mainly

down-graded track to Hereford, but after a brief halt there *Eire* made an equally exhilarating assault of Llanvihangel bank, followed by a brisk run to Pontypool Road.

That afternoon Pontypool Road depot hosted 43 locomotives, all steam except a lone diesel shunter; it was a similar situation at the two Newport sheds, Ebbw Junction and Pill, and again the only interlopers were a few diesel shunters. During the trip no fewer than 208 steam locomotives of 32 different classes were recorded in that enthusiast's notebook showing that despite encroaching modernisation there was still a great deal of steam at work on British Railways. Little did I realise, for I was that enthusiast, that just over five short hectic years later we would all be witnessing what was then regarded as the final death throes of steam on BR – by dint of authority the end of it all.

But was it? On 7 March 1981 I undertook a similar journey to that of 1963 as far as Newport, although it started from Wigan behind electric power to Crewe. There was again a Class 40 diesel to Shrewsbury, but surprisingly to some, a steam locomotive again took over the train, this time Stanier Princess Pacific No 6201 *Princess Elizabeth*, positively gleaming in maroon livery and proudly bearing the headboard Welsh Marches Express. The climb to Church Stretton was every bit as enchanting as on the first occasion, but this time a short stop gave passengers a chance to photograph and admire No 6201 before it continued to Hereford.

At the cathedral city *Princess Elizabeth* gave way to another Stanier product, Class 5 4-6-0 No 5000, and what memories were rekindled as its blast echoed around the Welsh borders while confidently lifting the packed train to Llanvihangel summit. Pontypool Road (now plain Pontypool) was passed in a hurry, the station reduced to a basic platform and shelter with no sign of any other activity in the area, quite unlike 1963. No 5000 spent nearly an hour at Newport rubbing shoulders with Inter-City

125s on Paddington–Swansea services before turning and watering, for working the train back to Hereford, where it retired to Bulmers Railway Centre.

For a time, following the banishment of steam from BR metals in the late 1960s, to have a steam trip such as I and many others enjoyed in 1981, and can still enjoy today, would have been said to be nothing more than a fantasy. That it is fact, not fantasy, is largely due to the plea, so vividly portrayed on the smokebox door of the engine on page 6, which has indeed been heard by many, as a result of which the iron horse is still very much alive and kicking!

This book describes how this has come about by tracing in words and pictures the run-down of steam in everyday service, both on BR and in industry, and the developments which have taken place over the past 30 years or so to ensure its continued existence. It is a remarkable story.

INTRODUCTION

Few inventions have had such a profound effect on life as Richard Trevithick's creation of 1804, creation being the operative word for the machine recognised by many as the nearest akin to nature. This, the first successful steam locomotive to run on rails, demonstrated by Trevithick along the Pen-y-darren Ironworks tramway, near Merthyr Tydfil, a year after what is now believed to be his first essay into a rail steam engine at Coalbrookdale in 1803, was a somewhat crude affair, and while quickly relegated to a stationary boiler, its successors were destined to revolutionise transport throughout the world. Indeed it was the combination of two technologies – the rail way, and the steam engine which provided the basis for expansion of the industrial revolution.

For two centuries previously, short sections of wooden rails, latterly with the addition of iron strips and towards the end of the period, iron rails, had been put down to ease the movement of waggons laden with minerals from mines and quarries to the nearest navigable waterway. Not surprisingly some progressive industrialists felt a desire for something more powerful than men or horses, and more controllable than gravity – the only available means of propulsion until Trevithick came on the scene.

After those initial trials by Trevithick in South Wales another eight years were to elapse before steam locomotives entered regular service – on the rack-rail Middleton Colliery Railway, Leeds, whose engineer considered adhesion alone insufficient. Meanwhile in the North East coalfields experiments continued, as a result of which a few engines began operating from various collieries, including George Stephenson's *Blücher* at Killing-

worth in 1814, the first to have flanged wheels. These developments led to the opening of the Stephenson engineered Stockton & Darlington Railway in 1825; while it was conceived as a coal carrying line, the inaugural train, proudly hauled by *Locomotion No 1* also conveyed passengers. The S&DR thus became the first public railway to use steam locomotives.

About this time a number of businessmen in South Lancashire were becoming increasingly unhappy with the transport facilities available between the port of Liverpool and Manchester, the centre of a growing textile trade, and some felt a railway would be beneficial. Following his earlier successes George Stephenson was appointed to build a railway; as completed the 31 mile route included some notable engineering feats, among them Olive Mount cutting, Sankey viaduct, and the crossing of Chat Moss bog – all still there today and carrying a railway over the same route.

After much deliberation and doubtless some persuasive argument from Stephenson, the directors of the Liverpool & Manchester Railway opted for locomotive haulage, except that stationary engines were to be used on the steep inclines through the Liverpool tunnels. This decision paved the way for the historic locomotive trials staged at Rainhill in October 1829, won by Robert Stephenson & Co of Newcastle-upon-Tyne with *Rocket*. The following year, amidst much acclaim, the L&MR was officially opened on 15 September, although the celebrations were tragically marred by the death of Liverpool MP William Huskisson after being struck by *Rocket* at Parkside.

Once the L&MR was operative the potential of the iron road became readily apparent, and by 1842 some 50 companies owned just short of 2,000 route miles of track; thereafter they snowballed, the next 30 years seeing the system expand to over 15,000 miles. By 1914 a further 5,000 miles had been laid, linking almost every town and city throughout the country by the various railway companies. There

The evocative words 'Please don't let me die' lovingly inscribed by persons unknown, together with the mournful face drawn on the smokebox door of former LMS Stanier Class 8F 2-8-0 No 48305 say it all. The engine was withdrawn from BR service in January 1968 and photographed languishing in Woodham Bros scrapyard, Barry, South Wales, on 22 September 1979.

One of the early steam locomotives fortunately saved for posterity, the much renowned Stephenson's Rocket of 1829, albeit in its modified form – the cylinders were originally set much higher (see photograph of replica on page 121). Rocket normally resides at the Science Museum, South Kensington, London, but was loaned to Merseyside County Museums for the Summer of 1980 for the 150th anniversary of the Liverpool & Manchester Railway, and is viewed against an appropriate backcloth in the City of Liverpool Museum transport gallery on 11 August 1980.

A sight typical of that which endeared the steam locomotive to the hearts of many people; former LMS Class 5 4-6-0 No 45013 raises the echoes at Greenholme on the climb to Shap summit in charge of a down ballast train, while at the rear a welcome helping hand is given by BR standard Class 4 4-6-0 No 75026 during a south-westerly gale in October 1967.

were exceptions though and some proposals never saw the light of day, while a number of competing duplicate routes were constructed, along with some which were uneconomic from the start.

Much of the development of Britain during the nineteenth century, both industrially and socially, was directly attributable to the railways, which quickly established themselves as the prime movers of freight, handling both new traffic as well as large tonnages won from the waterways. Railways also made it economic to open up many previously landlocked reserves of raw materials for transport to processing centres or direct to ports for shipment, and similarly they were able to distribute the finished products. Furthermore, people experi-

enced travel at a speed and convenience hitherto unknown. In essence this was only made possible by railways with steam motive power instead of the horse, the original basic concept of the steam locomotive seeing continual development over following years with different types designed to handle the various categories of traffic – slow stopping trains, fast expresses, heavy freights, and so on.

While the railways entered the twentieth century as the main carriers of both passengers and freight they have since faced many difficult years. Their profitability has been upset by changing traffic flows, the varying political climate, and ever increasing competition from other modes of transport, to say nothing of the traumas encountered during the two world wars; yet throughout they have proved an invaluable national asset.

When the state took over responsibility for the railway network in 1948 steam was still the predominant source of motive power except on a few lines where electric traction had taken over; diesel power was represented by a comparatively small number of shunting locomotives and a handful of experimental main line locomotives were in service

The elder statesman of preserved working steam locomotives Liverpool & Manchester Railway 0-4-2 Lion, *built 1838, comes face to face with the last steam locomotive constructed for BR in 1960, Class 9F 2-10-0 No 92220* Evening Star *at Didcot Railway Centre on 26 September 1981. Both engines have hauled trains on BR main lines in recent times.*

or planned. Yet 20 years on, steam locomotives on British Rail's much reduced 12,400 route miles of standard gauge track had come to an end, having been replaced by more modern forms of traction. Those that then remained in commercial service were used by industrial concerns, principally engaged in shunting duties and transferring wagons over relatively short distances to and from BR exchange sidings, but even here the steam locomotive was a dying species.

It was only right that over the years certain types, as they became life expired or redundant, should be set aside for posterity. A few notable locomotives had already been kept for preservation and others were earmarked, but until the 1950s little real attention was given to preservation; thus many important classes in the lineage of steam development have been lost for ever, all the more regrettable in the case of a few which were once destined for immortality but subsequently broken up years later. Both the LMS and the GWR were guilty in this respect and due to the latter only one hardly representative engine (*Tiny* now at Buckfastleigh, Devon) has survived from the broad gauge era which ended in May 1892.

Main line steam in the eighties. Rekindling memories of former glories on 12 December 1981, Bulleid West Country class Pacific No 34092 City of Wells *thunders through a bitterly cold snow-clad Bentham with a special which it had taken over at Carnforth, bedecked as in the days when steam worked the Golden Arrow Pullman between London and Dover, including the British and French flags.*

In the years between the grouping in 1923 and nationalisation in 1948 it was the LNER which showed the most enlightened attitude towards preservation, financing in 1928 the first museum, at York, solely devoted to railways. In 1948 these exhibits, together with other engines saved by the constituent railways, totalling 44 in all, became the property of the British Transport Commission, and in late 1960, as dusk began to loom over steam it was announced that a further 27 locomotives would join this select group. The list proved controversial to say the least, the main grouse being in respect of omissions.

To house adequately some of this collection, the BTC opened in 1963 the Clapham Museum of Transport (also catering for London Transport relics), while a number of engines were dispersed to provincial museums, and others placed in store.

11

There was soon to be a radical change in government policy which meant closure of both the old York and Clapham museums in the early 1970s but the majority of the railway items were then transferred to a new National Railway Museum at York, an outstation of the Science Museum under the direction of the Department of Education and Science, opened in September 1975.

Yet despite the expansion of railway displays including locomotives in numerous museums throughout Britain, and not just those specialising in railways, in reality they convey little of the atmosphere that surrounded the steam age. This aspect can only really be savoured when coal is being fed into a glowing firebox, converting water into steam inside the boiler and under controlled pressure used to power the engine. In this state the iron horse becomes alive, possessing a unique aura, its mood and character personified by its exhaust, both visibly and audibly.

For a time the ultimate sad fate of the steam locomotive seemed inevitable but many who relished the steam railway craved its continuance, a passion which manifested itself in, and then gave impetus to the infant railway preservation movement, the foundations for which were laid unsuspectingly in 1951. In that year a small group of volunteers accepted responsibility for the ailing narrow gauge Talyllyn Railway in Mid-Wales, staving off, almost at the last minute, what would otherwise have been certain closure for the line. Today the Talyllyn can be likened to the Stockton & Darlington or Liverpool & Manchester, for as the network expanded rapidly following the founding of the latter, so the number of privately-run preserved railways largely operating as tourist and leisure attractions has mushroomed during the past three decades, with steam the dominant form of traction.

The key to this resurgence is a large body of people who, like the Talyllyn pioneers, work on these private railways for sheer pleasure – the practical railway preservationists – men and women from widely varying social backgrounds united in the common pursuit of keeping rust from the rails and steam alive. They spend countless hours not only driving the engines and acting as signalmen, for which they must be trained and qualified to the highest standards, but also booking clerks and ticket collectors, and the many less glamorous but equally necessary jobs such as cleaning, fund raising, and the various administrative functions. Others devote years to restoring rolling stock, signals and buildings to their former glories.

From this dedication there are, including narrow gauge types, over 1,150 preserved steam locomotives in Britain today, a large percentage operable or intended to be in future. Working steam can be viewed at some 90 locations, ranging from lines up to 20 miles long to others with a few yards of track, some concerns having prospered to such an extent that a nucleus of full-time employees is needed, although the heavy dependence on voluntary labour remains. A selected number of engines are also allowed on specified BR routes, for BR soon rescinded its total ban on steam when it realised that nostalgia could be a paying proposition.

Some may well wonder why people do it? What motivates them to work long hours at tasks which can be both arduous and dirty, and in conditions which sometimes can only be described as primitive? Back in the days of everyday steam BR had difficulty in keeping staff as fitters, cleaners, and firemen working daily at a filthy and arduous job. Yet today there are those who willingly endure those same conditions, although not on the same scale, and not always on a daily basis, to keep steam alive. Few who have known the living presence of the steam locomotive have failed to be enraptured by its charm. Indeed the charisma of the iron horse is such that it has seemingly defied the laws of nature and today lives on with a vengeance.

Before looking in more detail at recent history I would like to pay my own tribute to all who have contributed to ensuring the survival of steam traction. We owe these enthusiasts an inestimable debt in saving these important parts of Britain's industrial heritage. A senior preservationist once remarked as the fire was being dropped at the end of a day's activity, his face radiant with joy, 'There's nothing like being with the engines'. I hope that through these pages something of that same joy will be conveyed to the reader. If anyone has not yet sampled the wares on offer at the private railways hopefully they too may be tempted to try inhaling a little steam from a real live steam locomotive in action.

1
THE RUN-DOWN
OF STEAM ON BR

As clocks throughout Britain chimed midnight on 31 December 1947 they were accompanied by the piercing whistles of scores of steam locomotives, heralding not just a new year but signifying, too, an end to their old identities, for at that instant the railways were taken over by the state. Amongst the myriad of assets inherited by the new born British Railways were just over 20,000 steam engines of some 400 different classes, many dating back to pre-grouping days, but generally not in the best of health – for, like the rest of the railway system, they were still suffering from the neglect and after effects of the second world war.

Many matters demanded the attention of the new regime, foremost among them their policy regarding motive power, but any fears at this stage for the future of steam were quickly dispelled by the announcement that a new range of standard steam classes was to be introduced, to incorporate the best features from the stock of the individual companies; meanwhile the workshops continued building to current patterns. To assist in making comparisons various categories of locomotives were exchanged between the British Railways Regions during Spring 1948 although only limited trials were possible with former GWR types because of their more ample proportions which severely restricted movements away from their home territory – a legacy from the broad gauge era.

Under the direction of former LMS man R. A. Riddles the first of the new standard series, Pacific No 70000 emerged from Crewe Works in January 1951, and since it was Festival of Britain year the locomotive was appropriately named *Britannia*. In all during the next decade 999 standard engines of 12 classes were built, ranging from 30 Class 2 2-6-2Ts for branch line work to the 251 highly successful Class 9F 2-10-0s intended principally for heavy freight duties. Gradually their presence was felt throughout the network, and while many railwaymen admired them, in places they received a very half-hearted welcome – especially on parts of

former GWR territory where old loyalties died hard.

With the publication of the BR Modernisation Plan in 1955 the long term future of steam suddenly looked bleak, for it advocated the gradual replacement of steam by diesel and electric power, later to be seen by some as the panacea for all BR's shortcomings. Even so the demise of steam seemed far off since new construction was to continue for a short time and the plan prophesied that in 1970 some 7,000 steam engines would still be in service. Indeed the following few years saw further developments to the steam stud, such as the rebuilding of Bulleid's Merchant Navy class Pacifics along with 60 of his light Pacifics, and improving the entire GWR King class 4-6-0s by fitting double-chimneys and self-cleaning smokeboxes, thus confirming BR's intended reliance on steam for some time to come.

Originally BR planned a systematic conversion by areas and routes to modern traction, but before this, since it had very little experience with mainline diesels, BR decided to test over a three year period 174 diesel locomotives to different specifications from various manufacturers, before placing further orders. However it was not to be, for by late 1956 the British Transport Commission had abandoned this ideal and the stock of diesel locomotives was soon multiplying at a rapid rate. Simultaneously electrification of some routes with heavy traffic began in earnest, such as the third-rail scheme to the Kent coast and various suburban schemes, while completion of the overhead 25kV ac system between Euston, Manchester and Liverpool was speeded up.

On 1 January 1960 BR owned 14,453 steam locomotives with new ones still coming off the assembly line, the last the celebrated Class 9F 2-10-0 No 92220 *Evening Star* released in March of that year by Swindon Works in Brunswick green livery, fully lined out, and complete with copper capped chimney in the best GWR tradition. This

Ex-WD Austerity 2-8-0 No 90625 trundles by Leeds (Holbeck) shed with a long rake of coal empties on 23 August 1966. To the left of the signalbox other steam locomotives can be seen on the depot but shades of things to come are evident as a Peak class diesel-electric locomotive noses into the picture on the far left.

'Jinty' Class 3F 0-6-0T No 47507 fusses about Chester General on 22 August 1966 while on station pilot duty. Notice the former LNWR lower quadrant signals on the right, some of which lasted into the 1970s though with replacement arms.

proved to be a watershed, for from then on the frontiers of steam were in retreat against a rapidly advancing tide of diesel and electric motive power, and helped in no way by the philosophies of Dr Beeching, during whose period as chairman of BR in the 1960s many lines and services were purged from the network. Throughout the 1960s it was a common sight to see processions of engines, some with many years of useful life left in them, being ignominiously dragged to the scrapyard as BR, with what at times appeared unseemly haste, endeavoured to be rid of steam as quickly as possible. As can be seen in Table 1 the peak year for withdrawals was 1962 when no fewer than 2,924 were condemned.

The first region where steam disappeared was the Western, although the original planned date of 3 January 1966 was thwarted at the last minute by the erstwhile Somerset & Dorset, the closure of which had to be delayed until 7 March 1966. The projected image of the total modern railway was however foiled for some time by steam's persistence in making forays across its borders, both from the

Stanier two-cylinder 2-6-4T No 42626 pulls away from Lostock Junction for Bolton with the final passenger train from Horwich, suitably adorned with headboard and wreath on Saturday 25 September 1965. Ceremonial last trains were an all too common occurrence during the 1960s as the passenger network dwindled.

Table 1
BR steam locomotives in capital stock.

1 January 1960	14,453
1 January 1961	13,244
1 January 1962	11,691
1 January 1963	8,767
1 January 1964	7,050
1 January 1965	4,973
1 January 1966	2,987
1 January 1967	1,689
1 January 1968	362
1 September 1968	3*

*Vale of Rheidol narrow gauge locomotives.

15

A number of Gresley A4 Pacifics spent their last days working from Aberdeen (Ferryhill) shed, having been displaced from their former habitats further south by diesel locomotives. No 60034 Lord Faringdon *is pictured ready to leave Aberdeen with the 13.30 to Glasgow on 14 June 1966, three months before withdrawal.* (Roger Siviter)

Gresley Class J38 0-6-0 No 65914 coasts tender first towards Thornton Yard with a mixed freight from the Cowdenbeath direction on 16 June 1966. (Roger Siviter)

The previous day Class J37 0-6-0 No 64620 crosses Montrose Basin with a short southbound goods. The class was introduced in 1914 by the North British Railway and passed to the LNER at the 1923 Grouping. (Roger Siviter)

Maunsell Lord Nelson class 4-6-0 No 30857 Lord Howe gets into its stride leaving Bournemouth Central with an up train for Waterloo on 8 September 1962. The engine, new in December 1928, was withdrawn a week or so later, and in October the class became extinct on BR.

Midlands and the Southern Region. But the die was cast and the Western was quickly followed by the Eastern, whose remaining serviceable locomotives were withdrawn from the end of the Winter 1966 timetable, although like the Western steam still infiltrated its territory. In fact it was a continuing feature right to the end that steam often penetrated so called no-go areas either for operating convenience or simply because no diesel was available to take over; on the latter duties steam locomo-

There was once a labyrinth of lines on the Isle of Wight but when this photograph was taken on 25 May 1966 only the Ryde–Shanklin section remained. Here Class O2 0-4-4T No 14 Fishbourne, fitted with Westinghouse brake equipment standardised on the Island, departs Ryde St John's Road with the 13.19 Ryde Pier Head–Shanklin. Steam services on the Island finished at the end of that year.

tives and their crews often put up some sparkling performances, almost like a defiant last stand.

During 1967 the Scottish, Southern, and North Eastern Regions (the latter having merged with the Eastern at the beginning of the year) followed suit, so that by January 1968 only the London Midland had any standard gauge steam engines in capital stock. They totalled 359 locomotives of seven classes and were dispersed between 13 sheds in the North West. Their duties mainly covered freight diagrams, but Stanier Class 5 4-6-0s could be seen regularly on a number of passenger trains, including the Belfast Boat Express between Manchester and Heysham, until May, and on a few others right up to the final curtain call.

The 1960s were often frenzied and somewhat emotional for the steam devotee as class after class

Drummond Class M7 0-4-4T No 30033 runs into Salisbury with empty stock on 6 September 1962. The design dates from 1897 and the class could be seen on a wide variety of duties from suburban and empty stock work in London to distant country branches.

The Castle class 4-6-0s were normally employed on express passenger turns, but on 27 May 1960 No 4095 Harlech Castle is seen engaged on a rather humdrum yet demanding duty piloting BR standard 9F 2-10-0 No 92217 into Newton Abbot with an up parcels train from the west. In steam days double-heading was often necessary over the arduous switchback route between Plymouth and Newton Abbot.

The highly individualistic Great Western was fond of 0-6-0 pannier tanks for shunting, branch line and trip working; No 3709, a member of the 5700 class introduced by Collett in 1929, rearranges a few wagons at Watery Road Goods Yard, Wrexham, on 22 August 1966.

2-6-2T No 4150 awaits its next assignment at Severn Tunnel Junction on 20 October 1964. The engine survived BR's steam holocaust and is now on the Severn Valley Railway.

of well-loved species became extinct on BR, and every effort was made to observe, photograph, and record it in its everyday environment while time remained. As the death knell drew nigh during the first half of 1968 hundreds converged regularly on the North West to witness the final chapter, a period when many enthusiast specials traversed the lines in the area, some utilising the sole surviving Britannia class 4-6-2 No 70013 *Oliver Cromwell*, retained specifically for this purpose. By July only three sheds, Carnforth, Lostock Hall (Preston) and Rose Grove (Burnley) had any steam rosters and from the first Monday in August they too were taken over by diesels. The last bastions of steam had fallen.

On the following Sunday, 11 August 1968, thousands paid homage along the route of BR's final commemorative special which ran from Liverpool to Carlisle and back via Manchester and Ais Gill, three Stanier Black 5 4-6-0s and *Oliver Cromwell* sharing the honours, with passengers paying the then princely sum of 15 guineas (£15.75)

for the privilege. That evening as the fires were dropped, as far as BR was concerned the last rites had been performed, and apart from an occasional trip with the privately preserved Gresley A3 4-6-2 No 4472 *Flying Scotsman* by courtesy of a previously negotiated contract, steam was dead, or almost dead.

In Mid-Wales the last of the few narrow gauge lines handed down to BR by the GWR still operated between Aberystwyth and Devil's Bridge during the summer months, its only power being three steam locomotives. The 1ft 11½in gauge Vale of Rheidol Railway may have been an anachronism in the modern world, but it survives yet and thus the final nail in the coffin has never been hammered home!

When new, the 55 Britannia Pacifics, power classification 7P6F, were distributed over five regions, but from 1964 all were concentrated on the London Midland; in the sunset of its life No 70021 Morning Star (a former Western Region engine and shorn of nameplates) storms up grade from Penrith towards Shap summit with a southbound freight during Autumn 1967.

With an Isle of Wight ferry on the right, Class 4 2-6-4T No 80019 starts away from Lymington Pier on the 5½ mile run to Brockenhurst in October 1966. The 80xxx series was the most powerful of the three standard tank classes introduced, and the Lymington line was BR's last steam worked branch line.

While the straight shed was the most common in some locations a roundhouse was favoured by some companies, locomotives being stabled on short lengths of track radiating from a central turntable; although not as compact as the former they had the advantage in that engines were always easily accessible with the minimum of shunting; the main disadvantage was that any damage to the turntable locked the engines in. The larger depots employed more than one turntable as at the former GWR Old Oak Common shed, London, which had four.

Here viewed across one of the twin turntables inside Leeds (Holbeck) shed on 23 August 1966 Stanier Jubilee class 4-6-0 No 45697 *Achilles* is flanked by two more Stanier products, Class 8F 2-8-0s Nos 48283 and 48158. The small oval 55A plates on the smokebox doors indicate that they were allocated to Holbeck. During the BR steam era each shed was given a similar numerical/alpha code (based on the former LMS system); sheds grouped together for administrative purposes had the same number but were distinguished by the suffix letter. The principal depot in a group took the letter A.

It was at the motive power depots – commonly called sheds – that the majority of servicing and light maintenance requirements were attended to. Just arrived at Aberdeen (Ferryhill) shed on 20 May 1964 are, left to right, BR standard 2-6-4T No 80055, Class J37 0-6-0 No 64620 and ex-War Department Austerity 2-8-0 No 90705.

Standing side by side at Exeter St David's on 25 May 1960 are Maunsell Class N 2-6-0 No 31838 with a freight for the Southern Region and recently introduced diesel-hydraulic Warship class No D811 Daring with the Paddington–Kingswear Torbay Express. Both locomotives have since been withdrawn, Daring after a life of only 12 years!

Rebuilt Bulleid Merchant Navy class 4-6-2 No 35008 Orient Line sojourns at Nine Elms depot, London, along with an unmodified West Country class Pacific in May 1967. Nine Elms, coded 70A, like many other sheds closed its doors with the steam age.

Glimpsed from a Windermere bound diesel multiple-unit near Milnthorpe on 22 July 1967 Britannia Pacific No 70010 Owen Glendower rushes by with an up passenger train. To be able to observe the road ahead in this manner was one bonus of the modernisation programme.

In many places the new diesels had to be serviced alongside steam – a far from ideal arrangement in that the former are best suited to a cleaner atmosphere than could generally be found at the smoke-laden steam sheds. At Perth depot Nos D5317 and D5120 (later Nos 26017 and 24120 respectively) bide time with former Caledonian Railway 0-4-4T No 55204 and Gresley V2 2-6-2 No 60970 on 20 May 1964.

The stark reality of BR policy. Already devoid of nameplates, coupling rods, and with no tender, Stanier Coronation Pacific No 46243 City of Lancaster looks forlorn as it awaits the cutter's torch at the Central Wagon Works, Ince, Wigan, on 1 July 1965. Ironically the yard lies within earshot of the West Coast main line, a route City of Lancaster knew so well in happier times.

The Southern Region main lines from Waterloo to Weymouth and Salisbury had the distinction of witnessing the final steam worked trains into and out of the capital, the last remnants succumbing on 9 July 1967, whereupon diesel and electric traction took complete control. With just over a year left on 23 May 1966 rebuilt Bulleid Battle of Britain class Pacific No 34090 Sir Eustace Missenden, Southern Railway roars through Micheldever with the 9.55 Bournemouth–Waterloo. The nameplates had already been removed from No 34090, an indignity suffered by many engines towards the end of their careers.

The Westmorland fells (now Cumbria) had long been a magnet for railway enthusiasts and with steam doing battle with Shap until the end of 1967 more and more were attracted to the locality. Although down trains on the West Coast main line face adverse gradients, apart from brief respites, from just beyond Carnforth at near sea level, the incline proper extends for five miles from Tebay, the final four miles to the summit, 916ft above sea level, being at 1 in 75; this often necessitated the use of banking engines in steam days. Summoning up every ounce of energy for the climb ahead in October 1967 Stanier Class 8F 2-8-0 No 48125 draws a train of vans across the Lune at the foot of the incline. The procedure for trains requiring help was to stop just beyond Tebay station while a banker ran out from the shed (just behind the photographer) and buffered up at the rear, before the ensemble, after the appropriate exchange of

whistle signals, attacked the gradient. BR standard Class 4 4-6-0 No 75026 came to the aid of No 48125 on this occasion. Today banking is but a memory at this remote but pleasant spot, electric trains speeding through almost as if the climb does not exist, while there is little evidence of the station or shed which once bustled with activity.

The end draws nigh. May 1968 and hard to comprehend that three months later scenes such as this would have gone forever. Here Stanier 8F 2-8-0 No 48445 is at grips with the 1 in 101 through the weed strewn platforms of the closed Hoghton station on the Preston–Blackburn line with a freight bound for East Lancashire. The crew seem to be enjoying the attention they received from the innumerable photographers who flocked to the area to record steam's twilight hours in BR everyday service.

2
REFLECTIONS OF THE LAST NIGHT

On Saturday, 3 August 1968 from eight o'clock onwards, railway enthusiasts from many parts of the British Isles began to assemble at Preston station. Two taxi-drivers remarked about the unusually long queue formed at the ticket office, oblivious of the real significance of the occasion. Mostly only two destinations were being booked – Blackpool South and Liverpool Exchange – by the last two ordinary-service trains on British Railways standard-gauge metals to be hauled by steam engines.

The 20.50 Preston–Blackpool train, seen on an earlier occasion to that described opposite, at Preston during Winter 1968 with Stanier Black 5 4-6-0 No 45149 in charge.

The platforms were a seething mass of humanity, sporting numerous cases carrying cameras and tape-recorders, while two bearded gentlemen clad in top hats and tails, and carrying a mock coffin draped with slogans on the impending demise of the steam locomotive, marched in sombre style past the Brush type 4 and coaches of the 17.05 from Euston, the rear portion of which was to form the 20.50 to Blackpool.

As Stanier Class 5 4-6-0 No 45212 backed on to this train, packed to capacity, photographers jostled with each other for suitable positions, and tape-recordists started the reels turning in an attempt to record her every beat and hiss. Soon No 45212 was gone, the last of a great pageant of steam

locomotives which have carried thousands to the Fylde Coast over the years.

Pending the expected arrival at 21.12 of the 17.25 from Glasgow, many people went to the other end of the station to photograph and admire another Stanier Class 5 No 45318, waiting to take the Liverpool portion. As the train came in, the tape-recordists rushed to get into the front coaches of the section due to depart at 21.25 for Liverpool, while photographers fortunate enough to possess flash stood ready to take their final shots of steam at Preston station.

With a blast from the exhaust, and to a battery of flash guns, No 45318 pulled her train gently out: this was the moment of no return. As she climbed steadily up the West Coast main line to Farington Curve Junction there were still isolated flashes of blue light from photographers at strategic vantage points on the lineside. Once over the short sharp climb to Moss Lane Junction, No 45318 began to hurry her train along into the night in the best traditions of steam (time-recordists noted an '80' at one point).

All too soon we were passing through the Liverpool suburbs, past Aintree where the outline of the now-silent engine shed, its coaling plant still towering into the sky, appeared a silent tribute to the golden age of steam. Three miles farther on hardly a trace could be seen of the site of Bank Hall shed, another former Mecca of the steam enthusiast – a stark reminder that an era was drawing to a close.

As No 45318 proudly brought her train slowly to a standstill in Liverpool Exchange the enthusiasts

The last day, Sunday 11 August 1968. Britannia Pacific No 70013 Oliver Cromwell *pauses amidst a throng of well-wishers at Ais Gill summit on its valedictory run from Manchester to Carlisle with the special which marked the demise of steam on BR. Surprisingly no commemorative headboard was carried, only the train reporting number 1T57. (Allan Heyes)*

gathered round to pay their last respects: the few ordinary passengers filed past with hardly a look at the engine, mildly surprised, partly amused at all the fuss. There was a chorus of "Auld Lang Syne" and three cheers for the steam locomotive.

There was a carnival atmosphere about the whole affair, yet it was tinged with sadness. There were still the six enthusiasts' specials to come on the next day and the very last run of all the following Sunday. There are steam visits to private lines and industrial systems to look forward to, but this was the last occasion on which one could pay one's fare at a British Railways booking-office window, and travel behind a steam locomotive on an ordinary service train.

★ ★ ★ ★

The above article was first published in the October 1968 edition of *Railway Magazine* under the pseudonym 'A Mourner', and is reprinted here by kind permission of the editor. The mourner was none other than the author, who is delighted to have been proved wrong in saying that it was the last occasion when you could purchase a ticket at a BR booking office for travel behind steam – albeit to date not on ordinary service trains!

3
LAST VESTIGES OF INDUSTRIAL STEAM

Since the pioneering days of railways, freight has been fed in and out of the main network along track owned by various mines, quarries, docks and factories, some also operating quite extensive internal railway systems. During the nineteenth century many industrialists soon realised the benefits of steam traction for hauling traffic between the point of loading or discharge and the railway companies' exchange sidings or, where the old order held sway, the canal staithes. A few even operated a passenger service but usually this was for the convenience of employees only.

Following the development of the tank locomotive in the 1840s they quickly came to dominate the industrial scene, since they proved highly suited for shunting and running either chimney or cab first, 0-4-0 and 0-6-0 wheel arrangements being the most popular. In the main they were purchased from such private locomotive building firms as W. G. Bagnall, Andrew Barclay and Hunslet, which had to look to this market, and overseas, for the sale of their products since most of the main line concerns, once established, developed their own workshops in order to serve their own needs. In fact some of the latter's engines were sold to the private lines after withdrawal, while a few industrialists built their own.

During the past 150 years, hundreds of these engines have fussed about industrial zones giving yeoman service, yet compared to their more illustrious main line counterparts few people paid them much attention. But as darkness settled on BR steam more and more enthusiasts began to search out the remaining industrial types still at work, a world many regretted not having explored earlier, for here, too, the iron horse was plummeting towards extinction, although by no means anywhere near eclipsed.

At the time of steam's death on BR in mid-1968 there were over 1300 standard gauge industrial steam engines dotted around Britain, over half in the hands of the National Coal Board and its subsidiaries. And what a diverse group, the craftsmanship of over three dozen factories being represented, while about 50 dated from the nineteenth century, and a few were ex-BR locomotives. The largest class consisted of over 200 Hunslet-designed WD Austerity 0-6-0STs chiefly owned by the NCB and originally built for service during World War 2, although construction continued long after the end of hostilities, the last in 1964 when Hunslet finished the 484th example. Six other firms laid the frames for 267 of them. As Class J94 75 Austerities saw service with the LNER and BR.

In 1968 nearly 50 fireless locomotives were also extant. These locomotives, as their description shows, had no firebox, and in their day's work were charged periodically with steam from a stationary boiler. They were used in highly inflammable zones such as chemical works where it would be dangerous to use conventional engines. Operating as well was Bowater's United Kingdom Pulp & Paper Mills Ltd 2ft 6in gauge railway at Sittingbourne, Kent, the last remnant of once numerous industrial narrow gauge steam lines.

However in a sense the statistics belie the true picture, for dozens of these engines had already in effect been discarded and simply awaited disposal;

An example of an ex-BR steam locomotive in commercial service after the fateful date, 11 August 1968. In NCB ownership former GWR 0-6-0PT No 7754, withdrawn by BR in December 1958 but with its smokebox number plate still in place, waits impatiently at Mountain Ash, South Wales, ready to bank a heavy diesel-hauled train for Aberaman Phurnacite Plant, Abercwmboi, on 5 October 1971. In September 1980 No 7754 embarked on a third stage of a distinguished career when it was moved to the Llangollen Railway, North Wales, where hopefully it will see further use in due course.

With a cataclysm of sound two Andrew Barclay 0-6-0STs, works Nos 1175 of 1909 leading and 2358 of 1954, NCB Nos 8 and 25 respectively, storm up to Polkemmet Moor with coal from Polkemmet Colliery, Whitburn, on 23 May 1974. As a monument to its illustrious past No 8 today stands on a plinth by the colliery entrance – albeit carrying the wrong builder's plates!

Under a stormy sky at Dunaskin Washery, Waterside,
Ayrshire, NCB No 19 Andrew Barclay 0-4-0ST No 1614, built
in 1918, prepares to push coal from Pennyvenie Colliery, some
3½ miles distant, to the back of the washery on 30 August
1973. Through the good offices of the NCB many engines have
been spared the breaker's hammer on withdrawal, No 19 being
loaned to the Dalmellington Countryside Society and now
awaiting restoration at the site of Minnivey Colliery, one of its
old haunts on the Waterside system.

In contrast to their NCB counterparts a quartet of four-wheeled
vertical-boilered Sentinels soldier on at R. B. Tennent Ltd,
Whifflet Foundry, Coatbridge, and on 18 June 1980 the
youngest, Denis built 1958 is seen at work.

Traffic is halted at Burradon while NCB No 48 (Hunslet Austerity No 2864 of 1943) drifts across the main street returning to Eccles Colliery, Backworth, from Weetslade Central Coal Preparation Plant, just north of the Tyne, on 28 August 1974.

many others were retained for use only when a diesel was out of action. By the dawn of the 1970s the steam ranks had been reduced by a further third, including Bowater's narrow gauge engines although under a timely preservation scheme part of the line has continued to run utilising much of the same stock, though purely as a passenger pleasure line.

During the 1970s the decimation of steam continued apace, some replaced by diesels, others through the introduction of road haulage or conveyor belt systems. Colliery and factory closures and rationalisations have accounted for the loss of some engines, while the extension of the merry-go-round principle, first introduced in the mid-1960s, whereby wagons are automatically loaded and discharged at slow-speeds without uncoupling and with the BR locomotive attached to the train, has also taken its toll.

From all these reasons it has become progressively more difficult to come across steam in commercial use, and by 1976 numbers had dwindled to about 250, of which 60 per cent were on the books of the NCB. Meanwhile many enthusiasts made the most of the remaining steam activity, and some locations, like the NCB systems at Mountain Ash, South Wales, and Waterside, Ayrshire, almost received entries in the tourist guides because of the number of visitors! Realising the interest in their locomotives some operators have considerably organised open days and tours of their railways, the latter usually in open wagons or brakevans, events which all too often in recent times sadly marked the end of steam, as at the CEGB power stations at Acton Lane, London, and Agecroft, Manchester, during 1981.

At the time of writing 61 steam locomotives nominally remain in industrial use, but there are barely a handful of locations (including those which employ fireless locomotives) where regular running occurs and while some hopefully await the call to

A line-up of no fewer than five engines inside Backworth shed on 27 August 1974. Odd man out among a group of Austerities, second left, is No 16, an outside-cylindered 0-6-0ST built by Robert Stephenson & Hawthorns in 1957, works No 7944; the other three Austerities keeping No 48 company are No 49, another RSH No 7098 of 1944, a second No 48, Hunslet No 3172 of 1944, and No 6, a Bagnall example No 2749 of 1944. Little trace now remains of the once busy Backworth railway complex or the mines it served, but Nos 16 and 49 survive on the Tanfield Railway, situated on the other side of the Tyne, and Hunslet No 2864 on the Strathspey Railway, Scotland.

Following the closure of Bedlay Colliery, Glenboig, in December 1981, Bold Colliery, St Helens, became the final outpost of regular steam on the NCB; sadly this ended during Autumn 1982 when a diesel arrived to take charge. On 20 August 1974 Whiston, a Hunslet built Austerity No 3694 of 1950 stands ready to heave two BR merry-go-round wagons into the exchange sidings for despatch to the CEGB Fiddler's Ferry Power Station, near Widnes, where the load will be automatically discharged into the storage bunkers as a BR Class 47 diesel controls the train at slow speed. Unlike Bold many collieries have been equipped with overhead loading bunkers whereby this type of wagon can similarly be filled without the need to detach the BR locomotive, this highly efficient way of transferring coal from the pits to the power stations being responsible for making many industrial locomotives (diesel as well as steam) redundant.

duty should the need arise, many will never turn their wheels again in revenue earning service at their present abodes. Fortunately, as is documented later, not all the redundant engines have been consigned to the scrap-metal merchants, some taking on a new lease of life at preservation centres.

But in the words of the proverb 'while there is life there is hope' – and so it would appear, for of late there has been some ironic twists to the tale. To the delight of steam devotees on more than one occasion commercial concerns have seen fit to hire locomotives from preservationists for short periods to cover for a recalcitrant diesel, while in 1974 scrap dealers Crossley Brothers (Shipley) Ltd, near Bradford, purchased Andrew Barclay 0-4-0ST No 1823 of 1924 from the Yorkshire Dales Railway as a stand-by to their diesels, and it has recently been fully overhauled. Perhaps it should be mentioned that employed at Crossley's are one or two steam enthusiasts, the reason for the longevity of a number of other locomotives.

Another unusual happening occurred in West Yorkshire during Autumn 1981 when Hunslet Engine Co Ltd carried out various experiments with different grades of coal using an NCB Austerity 0-6-0ST fitted with their underfeed stoker and gas producer system, first at Allerton Bywater Colliery and then based at Wheldale Colliery, Castleford. The locomotive saw further service on conclusion of the tests.

The sands of time may be running out for industrial steam but with the occasional, if at times unexpected spark of life still evident it could be many years yet before the final death sentence is pronounced. May it be so.

In recent times the only firm to employ a steam locomotive on the Trafford Park industrial estate, Manchester, was cornflour manufacturer CPC (United Kingdom) Ltd; on 22 September 1977 Andrew Barclay 0-4-0ST No 1964, built 1929, on its daily foray outside the works entrance, moves a string of coal empties to the nearby sidings for transfer to BR by a Manchester Ship Canal Railway diesel. Discarded shortly afterwards No 1964 can be viewed at the Greater Manchester Museum of Science & Industry, Liverpool Road, Manchester.

Hunslet 0-6-0ST No 3715 Primrose No 2, built 1952, takes a breather in the evening sunshine on the tip at Peckfield Colliery, Micklefield, while colliery waste is deposited from some side-tipping wagons on 24 August 1972. Primrose No 2 later transferred its affections to the Yorkshire Dales Railway, Embsay, near Skipton.

Coal mining has for many years been a staple industry of South Wales, steam playing a vital role in moving the black diamond out of the Valleys. In the next valley to where Trevithick's 1804 Pen-y-darren locomotive first saw the light of day Andrew Barclay 0-6-0ST No 2074 Llantanam Abbey, built 1939, is photographed against a powerful sun 175 years later as it moves coal destined for Aberaman Phurnacite Plant, Abercwmboi, away from Mountain Ash, near Aberdare, on 21 September 1979. Before being taken over by the NCB this section of track formed part of the ex-GWR Pontypool to Neath route, from which passenger services were withdrawn on 15 June 1964. The regular daily steam workings from Mountain Ash shed came to a sudden unscheduled end in December 1979 when flood water caused extensive damage to a bridge over the River Cynon which effectively cut the system in two, forcing the NCB to reorganise its rail traffic, since when there has been very little steam activity.

The coalfields of North Wales are more localised and by no means as extensive as those in the south, the last to use steam being Bersham Colliery, Rhostyllen, near Wrexham. On 17 September 1979 0-4-0ST *Shakespeare*, of pensionable age but looking strong and healthy draws coal away from the screens; bottom right are the tracks of the BR Chester–Shrewsbury line. *Shakespeare*, built by Hawthorn Leslie in 1914, works No 3072, was retired shortly afterwards.

Displaced by diesels but in exemplary condition 0-4-0ST twins CEGB 2 and CEGB 1, both built in 1954 at the Newcastle-upon-Tyne works of Robert Stephenson & Hawthorns (Nos 7818 and 7817 respectively) are ready for action, in case of need, inside the shed at Castle Donington Power Station, Leicestershire, on 22 August 1980. Occasional open days have been held with the engines in steam.

At work in the Kent coalfield, Britain's smallest, 1954-built Hunslet Austerity 0-6-0ST No 3825 eases into the exchange sidings at Snowdown Colliery, Nonington, on 16 September 1974. To the left of the signalbox is the BR London Victoria to Dover via Canterbury line, while the overhead wires were for the use of electric locomotives when visiting the sidings; they took current from the 750V dc third rail when on the main line.

For just one day steam made a welcome return to Birkenhead Docks on 22 July 1978 when Avonside 0-6-0ST Lucy, works No 1568 of 1909, owned by the Liverpool Locomotive Preservation Group, made a tour of the dock system. Here Lucy edges its train beneath three dockside cranes, the four brakevans and three mineral wagons full to capacity with visiting enthusiasts; on the right is one of the resident diesel shunters. Lucy was taking time-out from its home at Steamport Transport Museum, Southport. For many the lifebelt on the left was a reassuring sight!

English China Clay's small Bagnall 0-4-0ST No 3058 Alfred, built 1953, is positively dwarfed by its train at Par, Cornwall, on 15 June 1977; the squat profile had been necessitated by the need to negotiate a low bridge in earlier years. Today Alfred is cared for by the Cornish Steam Preservation Society at Bugle, near St Austell.

4
NARROW GAUGE INSPIRATION

In Britain, compared with many countries overseas, only a comparatively small number of public railways were built using tracks with a gauge less than the 4ft 8½in adopted as standard for the main system, although many industrial concerns, such as

mines, quarries etc, have put the attributes of the narrow gauge to good use. Despite their sparsity, and in some cases perhaps because of, many of the public narrow gauge lines that did operate found life very difficult for a variety of reasons, and by the start of the second half of the twentieth century only a few remained in business.

One that did linger was the 2ft 3in gauge Talyllyn Railway on the West Wales coast, but in July 1950, following the death of the owner Sir Henry Haydn Jones, the line looked set, like so many of its contemporaries, to fade into oblivion. The Talyllyn first opened in December 1866, its main purpose to carry slate from Bryn Eglwys quarry to Tywyn for onward transmission by the Cambrian Railway, a passenger service also being provided. The quarry and railway came into Sir Haydn's ownership in 1911.

When the quarry ceased production in 1946 and following clearance of the remaining stocks, in deference to Sir Haydn's wishes the railway stumbled on, using the same original equipment to provide a limited service for the local inhabitants and visiting holidaymakers. Shortly after his death Sir Haydn's executors were poised to wind-up the by then very run-down railway, and this would certainly have been its fate had not a group, mainly from the Midlands, heard of its plight and, in the nick of time, formed the Talyllyn Railway Preservation Society with the avowed intention of securing the future of trains along the Afon Fathew Valley. They knew little of what lay ahead and between them had only a negligible amount of practical railway operating experience, but, despite

The doyen of them all – the Talyllyn Railway. On 30 May 1979 Kerr Stuart 0-4-2ST No 4 Edward Thomas *slowly crosses Dolgoch viaduct with a Tywyn-Nant Gwernol train. Built in 1921* Edward Thomas *spent its working life on the Corris Railway before transferring its allegiance to the Talyllyn in 1951, one of four steam engines which have supplemented the original two, and needed to handle the growing traffic generated by the Talyllyn RPS.*

many difficulties, they succeeded, trains first running under the new regime on Whit Monday 1951.

The initial objective achieved, the founding fathers of the Talyllyn RPS were not content simply to rest on their laurels and allow the 6½ mile line to totter along as before, almost as if held together by a shoe-string, and through sheer determination and hard work the tumbledown state was gradually transformed into the well maintained and efficient railway it is today. A high spot was reached in May 1976 when a ¾ mile extension was opened from Abergynolwyn to Nant Gwernol, the point where the slate from the quarry used to be collected. Over the years additional stock, both locomotives and coaches, have had to be obtained to cater at times for a seemingly ever increasing demand, 118,665 passengers being carried in 1982, against 15,628 in 1951.

As the new management at Tywyn was finding its feet, another group became active 30 miles north up the coast at Porthmadog – its aim to revive a second former slate carrying line, the 1ft 11½in gauge Festiniog Railway, which had been aban-

doned in 1946. In July 1955 it too tasted success when trains returned to the Traeth Mawr embankment (the Cob) between Porthmadog and Boston Lodge, since when the line has been extended in stages until in May 1982 trains started running over the full 13½ miles through to Blaenau Ffestiniog, where the FR shares a new station with BR's Conwy Valley services from Llandudno. During the intervening years many obstacles had to be overcome, chief among them a reservoir built to serve the hydro-electric Ffestiniog Power Station which flooded part of the original route above Dduallt, necessitating the construction of an entirely new formation at a higher level, including

Festiniog Railway 1893-built Hunslet 2-4-0STT Blanche *heads away from Porthmadog across the Cob with the 19.00 to Tanygrisiau on 29 May 1979. In 1955 the railway over the Cob was the first section of the revitalised Festiniog to be reopened following nine years of inactivity, another 27 elapsing before the ultimate goal of Blaenau Ffestiniog was reached. It was no mean feat considering the tremendous difficulties encountered during the intervening years.* Blanche *in its original form as an 0-4-0ST was obtained from the closed Penrhyn Quarry Railway, being rebuilt by the Festiniog which also converted it to an oil-burner, along with the rest of its stud.*

what is Britain's only spiral on a statutory railway.

Throughout the annals of history many men have unwittingly been the instigators of great things, and such was the case with those who helped rekindle the ethos of the Talyllyn and Festiniog railways, today both major tourist attractions. Little did they realise that from those first tentative resolute steps would grow a preservation movement which now extends to virtually every corner of Britain and embraces many gauges.

Latterly among the preserved railways the standard gauge has predominated and with which this volume is primarily concerned, but undoubtedly the inspiration for all that has been achieved initially came from the early narrow gauge revivals. There are few household names listed among the pioneering preservationists: exceptions are author L. T. C. Rolt at Tywyn and Alan Pegler, later of *Flying Scotsman* fame, at Porthmadog, but while as individuals they are largely unknown they nevertheless deserve honour. It was they who proved conclusively that working steam preservation was a practical proposition, and thus paved the way for an ever growing army of second generation railway builders.

Britain's only rack-rail line, the 2ft 7½in (800mm) gauge Snowdon Mountain Railway which climbs almost to the summit of Wales's highest mountain (3560ft), has seen continuous operation except for the winter months since 1897, following a catastrophic opening day the previous year. With Llanberis Lake in the valley seen above the engine, one of the seven 0-4-2STs which handle all the traffic, No 7 Aylwin (now named Ralph Sadler) pushes its one coach up the lower slopes on 12 September 1978. The engines were built by Schweizerische Lokomotiv & Maschinenfabrik, Winterthur, Switzerland, between 1895 and 1923, and take their drive through the rack pinions.

British Rail's only steam survivor – the 11¾ mile 1ft 11½in gauge Vale of Rheidol Railway. On a dull 15 September 1978 one of the three 2-6-2Ts which monopolise the line, No 8 Llywelyn leaves Aberffrwd after taking water with the 14.15 Aberystwyth–Devil's Bridge.

44

The 15in gauge Fairbourne Railway has already celebrated its diamond jubilee and on 14 September 1978 2-4-2 Sian, built in 1963 by Guest's of Stourbridge, skirts the southern flank of the Mawddach estuary at the approach to Barmouth Ferry with a train from Fairbourne. On this blustery afternoon the majority of passengers deemed it best to travel in the enclosed coaches!

Since the late 1920s the 15in gauge Romney, Hythe & Dymchurch Railway has been an integral part of the Kent coastline and in every sense a main line in miniature. On 14 June 1978 4-8-2 No 6 Samson, built for the line by Davey Paxman in 1926, leaves Dymchurch with the 14.20 Hythe–New Romney and passes one of the modern colour-light signals which control sections of the route.

Inspiration in the 1950s came, too, from a select band of narrow gauge steam lines operating commercially but looking principally to tourists for their revenue, and with no thoughts of changing to any other form of traction, lines which are still active today. In this category come the rack-rail Snowdon Mountain Railway, the 15in gauge 13¾ mile Romney, Hythe & Dymchurch Railway and BR's last remaining all-steam line, the Vale of Rheidol Railway from Aberystwyth to Devil's Bridge. These lines also demonstrated that this kind of self-contained railway could be viable.

The full story of the steam-operated narrow gauge railways in modern times is narrated in my book *Narrow Gauge into the Eighties* published by David & Charles, but it is appropriate here to take a further brief look at those concerns which have had so much influence on recent railway history.

The Ravenglass & Eskdale Railway, familiarly known as the 'Ratty', was first opened in 1875 as a 3ft gauge line, being converted to 15in in 1915. Threading glorious Lakeland shortly after departure from Irton Road with the 16.40 Dalegarth to Ravenglass is R&ER 1976-built 2-6-2 Northern Rock on 22 July 1979.

5
STANDARD GAUGE PIONEERS

To revive and administer a narrow gauge railway was one thing; to attempt the same with a standard gauge line presented a whole new challenge. Theory and practice may well be similar but the element of scale makes the latter a totally different proposition, and in the first instance the mere contemplation of operating a 4ft 8½in gauge railway in the manner of the Talyllyn or Festiniog must have appeared somewhat daunting.

However for the majority of railway enthusiasts

Ambassador for steam – Alan Bloom. A horticulturist by profession, Alan Bloom's love of steam has manifested itself in a fine museum within his grounds at Bressingham Hall, Diss, Norfolk, which features not only locomotives but such other aspects of steam as traction engines, roundabouts, etc. He is portrayed on the footplate of 1ft 11½in gauge Hunslet 0-4-0ST Gwynedd *of 1883 which arrived from Penrhyn Quarries, North Wales, in 1966, the standard gauge stock beginning to accumulate from the following year. In 1968 Bressingham took delivery of London, Tilbury & Southend Railway 4-4-2T No 80* Thundersley, *and Britannia class 4-6-2 No 70013* Oliver Cromwell *immediately after BR's farewell to steam special on 11 August, both on loan from the National collection. Since those days the museum has expanded steadily.*

in the 1950s and 1960s it was the engines owned by British Railways which were dearest to their hearts. It was these locomotives which were observed for countless hours from the end of station platforms, or from suitable bridges and embankments, many also travelling the length and breadth of Britain in search of engines not normally located in their home areas, at the same time often visiting those never to be replaced institutions – the engine sheds. After first sighting an engine, its number (and name) was meticulously underlined in a cherished copy of the latest Ian Allan *abc* and oh, what joy and sense of satisfaction when the last member of a particular class that was needed to complete a set was spotted, sometimes far away from its regular routes.

During this period the steam railway and the pursuit of locomotives was for many a way of life; its magnetism became abiding, and little if anything could replace it. In the past when progress or wear and tear had dictated that favourite classes of engines be retired, they had in the main been superseded by more modern versions which in turn had won the affections of the enthusiasts. Generally steam had replaced steam.

This time it was to be so different, for when the last batch of BR standard Class 9F 2-10-0s had taken to the rails from Swindon Works (the last was released in March 1960) there were to be no more.

Example of an engine purchased by a society. When Gresley A4 Pacific No 60007 Sir Nigel Gresley *was withdrawn by BR in February 1966 it was bought by the A4 Locomotive Society and subsequently despatched to Crewe Works for overhaul, from where it emerged sporting its original Garter blue livery and number. As No 4498 the 100th Pacific built to Gresley's designs powered a number of specials during 1967 before the tracks were cruelly cut from under its flanged wheels by BR's steam ban, happily now long since relaxed as viewed here near Giggleswick on 23 May 1981. No fewer than six A4s were reprieved during the 1960s although two, Nos 60008* Dwight D Eisenhower *and 60010* Dominion of Canada *were shipped across the Atlantic to USA and Canadian Museums respectively.*

The very future of live steam was in jeopardy. Taking their place were what then appeared to be characterless diesels and equally characterless, if not more so, electrics. The seemingly unfeeling bureaucratic decisions had been made: steam as known and admired by thousands on the main and branch lines of BR was due to die. At BR headquarters and in the appropriate corridors of transport power in government there was no place for sentiment or nostalgia; it was only a matter of time before the policy was fully enacted, and in some circles the sooner the better. There would be no reprieve.

As outlined in the introduction a selection of varying types was destined for official preservation; praiseworthy though this was, for anyone with steam in the blood there was no wish to see the iron horse simply reposing like the stuffed carcass of a wild beast in a natural history museum. With the inevitable end drawing ever closer, and being hastened by the drastic pruning of the system and the curtailment of many services, both goods and passenger, during the 1960s, it soon became apparent that if steam was to be seen at work in the future action was required urgently, not only in saving suitable examples but also in finding somewhere for them to run.

The precursor – Capt W G Smith's Ivatt designed 0-6-0ST No 1247 (BR No 68846), 17 years on in private ownership on 13 June 1976, leaves Goathland for Pickering with a North Yorkshire Moors Railway working. No 1247 is now at the National Railway Museum, York.

One of the most famous steam locomotives of all time – Gresley A3 Pacific No 4472 (BR No 60103) Flying Scotsman runs light through Manchester Victoria on 12 March 1980, such scenes being due entirely to the foresight of Alan Pegler who acquired the engine from BR early in 1963. Latterly the engine has been owned by another individual, Mr W H McAlpine.

Far from its native heath on the GWR Cambrian lines, *Dukedog class 4-4-0 No 3217* Earl of Berkeley *(BR No 9017 when withdrawn from Machynlleth shed in October 1960) draws forward at Sheffield Park, before hauling its two coaches to Horsted Keynes on the pioneer Bluebell Railway, deep in what was once Southern Railway territory, during July 1970. It is now common to find engines like* Earl of Berkeley *active on one-time alien tracks, for in the preservation era old boundaries have counted for nothing!*

Keighley & Worth Valley Railway reopening day 29 June 1968. The joyous scene at Oxenhope following the arrival of Ivatt LMS 2-6-2T No 41241 and (barely visible) USA 0-6-0T No 72 (BR No 30072) with the inaugural special from Keighley. The day marked the culmination of six years' endeavour to bring steam back to this West Yorkshire branch which was forsaken by BR in 1962. (Gordon Blears)

In 1959 a certain Capt W. G. Smith gave a positive lead when he purchased ex-Great Northern Railway Class J52 0-6-0ST No 68846 – the first engine bought by an individual from BR. A few others were in the happy position of being able to do likewise, and through their foresight a number of engines were saved from an undignified end at the hands of an unsympathetic scrap-metal merchant, among them the renowned *Flying Scotsman*, purchased by Alan Pegler in 1963.

Individually, not many, whatever their fancy, could afford the price demanded by BR for its unwanted assets, but various groups were formed in order to raise the required capital, some advertising widely for donations as well as organising the usual well-tried methods of fund-raising, although large loans sometimes had perforce to be negotiated if BR deadlines were to be met. Inevitably not all were successful in what became a race against time in a bid to save as much as possible. Yet by 1969 over 100 locomotives had been spared from BR's mass slaughter through private purchase – a most commendable achievement. Of these a small proportion were intended for static display only, such as the eight bought for exhibition at four Butlin's holiday camps, although some of these have now joined the active list.

In respect of actual operation the pattern was set for the standard gauge in August 1960 when the Bluebell Railway in Sussex, run entirely by volunteers, steam hauled its first public passenger trains from Sheffield Park as far as Bluebell Halt, just short of Horsted Keynes, the latter then still in use by BR. The Bluebell RPS, which at that time utilised this section of the former East Grinstead to Lewes line under a lease from BR, was only formed in March 1959 and it is remarkable how the initial enthusiasm for the project was so speedily turned into reality, no doubt spurred on by steam's hasty retreat from the South East.

At the beginning of 1960 the Middleton Railway Preservation Society was founded with a view to ensuring the future of this historic line following its abandonment by the National Coal Board, the impetus coming from Leeds University Railway

In the new era, trains on the Dart Valley Railway in South Devon were worked at first push-pull fashion, initially because of a lack of run-round facilities at the Totnes end, the operating section being severed short of the junction with BR's West of England main line. With one such train in the DVR's second year on 8 July 1970, GWR 0-6-0PT No 6412 is sandwiched between four GWR auto-trailers forming the 11.25 Buckfastleigh–Totnes. No 6412 has since transferred its affections to the West Somerset Railway. (Terry Flinders)

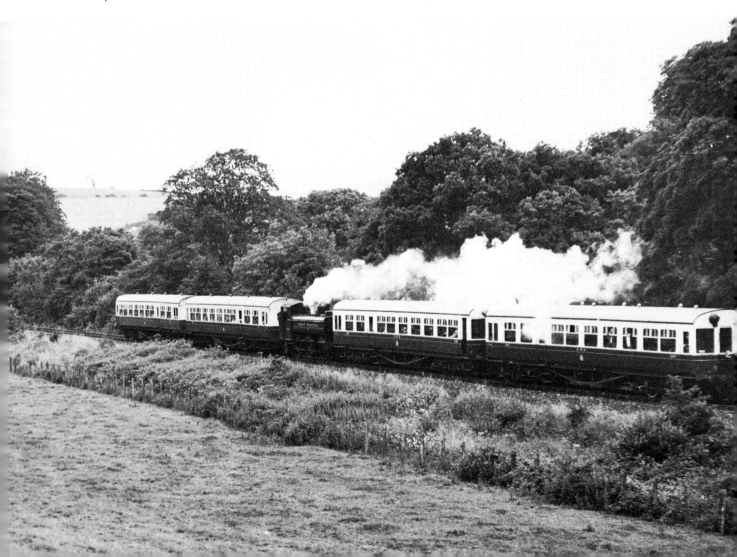

Society. Its pedigree deserved such attention since in 1758 it was the first railway to be sanctioned by Act of Parliament, as well as, in 1812, the first to use steam locomotives regularly. The Middleton RPS did in fact commence business two months before the Bluebell but the early passenger trains were diesel hauled, and the primary aim in those days was to serve local industry by conveying freight to and from the BR exchange sidings at Hunslet.

Encouraged by events at Middleton and Sheffield Park, along with the narrow gauge revivals, others soon had designs for yet more schemes. At the time few probably realised the full implications of what they were embarking upon, the legal complexities to overcome, and the patience required when dealing with government and local authorities, not to mention the vast amount of hard work involved, both manually and administratively.

Despite the many obstacles and frustrations, and the counter attraction of steam at work on parts of BR for much of the time, by the end of the 1960s another four concerns had inaugurated steam passenger services. Significantly three started as the last embers of fire on BR faded in Lancashire during 1968, these being the Lochty Private Railway, Fifeshire, (this had seen light engine movements only the previous year), the Keighley & Worth Valley Railway in West Yorkshire and the Foxfield Light Railway, Staffordshire, a former colliery line. The last to open its doors to the public during this period was the Dart Valley Railway in April 1969 with headquarters at Buckfastleigh, Devon, on the former Ashburton branch from Totnes, ironically officially re-opened by Lord Beeching during whose tenure as Chairman of British Rail many such lines, including that to Ashburton, had been axed from the system. The DVR, which aimed to recreate the atmosphere of a typical GWR country branch line, was the first to be organised on a commercial basis, although like the others there is a reliance on voluntary labour for its viability.

Some less ambitious schemes (at the time) were also set in motion. These took the form of live steam centres where periodically engines could be demonstrated along short sections of track, sometimes hauling an odd carriage or brakevan, while in

some cases footplate rides became the order of the day. Their number included the Great Western Society depot at Didcot, Steamtown Railway Museum, Carnforth, the Standard Gauge Steam Trust, Tyseley, later to become the Birmingham Railway Museum, and Dinting Railway Centre, near Glossop, Derbyshire.

By the turn of the decade plans for the reinvigoration of other lines were well advanced, such as those along the valley of the Severn from Bridgnorth, Shropshire, and across the North Yorkshire Moors from Grosmont, near Whitby, while some schemes were at an embryo stage. Not all achieved their desired aims, like the plans for the Yatton to Clevedon line and the Waverley route which never surfaced from the drawing boards, while one or two with ambitious ideas had to lower their sights and be content with a truncated scheme, as for example the Main Line Steam Trust based at Loughborough. Even so, firm overall foundations were put down by the steam preservation movement during the 1960s; the question that then remained to be answered was – how well would the seeds develop?

The roots of the Birmingham Railway Museum, then known as the Standard Gauge Steam Trust, sprang-up in 1969 and was based at an abandoned BR steam shed, the former GWR Tyseley depot. An early arrival was Stanier Jubilee 4-6-0 No 5593 Kolhapur, seen here with visitors queuing to take a closer look at the footplate on 4 April 1982.

One that failed. The preservation scheme intended for the Longmoor Military Railway, Liss, Hampshire, had a great deal of potential but regrettably came to nought, folding up in 1969, following which the engines were dispersed to other locations. Receiving the admiration of visitors at an open day on 28 September 1968 are Ivatt Class 2 2-6-2T No 41298 and Bulleid West Country Pacific No 34023 Blackmore Vale. *No 41298 is now at the Quainton Railway Centre, near Aylesbury, while* Blackmore Vale *is on the Bluebell Railway as No 21C123.*

(Gordon Blears)

54

6
SCRAPYARD WITH A DIFFERENCE

Traditionally the cutting-up of withdrawn BR locomotives was carried out within the confines of the main works, but during the late 1950s, as the modernisation programme gathered momentum, the engines awaiting the oxy-acetylene torch began to outstrip the facilities available for dealing with them, and a large back-log developed. In early 1959, with the problem becoming acute, BR decided to sell some of its redundant stock by tender to outside contractors; thus during the next few years the yards of various scrap-metal merchants became the final resting places for hundreds of unwanted steam engines.

Among the early buyers, if not the first in February 1959, was Woodham Bros of Barry Docks, South Wales, which purchased a batch of five ex-GWR engines from Swindon Works; with little delay they were reduced to small manageable pieces for resale to the steel industry. Like Woodhams, other merchants bidding in this market were also naturally very anxious to turn their acquisitions into ready cash as quickly as possible, and few engines lingered intact for more than a few months once they had arrived at their designated execution places.

However at Woodhams events took a very different course, for while other engines followed the first consignment at regular intervals, cutting-up was carried out at a much slower rate, and by the end of 1965 no fewer than 150 were standing idle around the yard. Astonishingly the dismantling of locomotives was then stopped altogether while the firm concentrated on other work, including large quantities of withdrawn BR freight wagons. Nevertheless, and perhaps surprisingly, Woodhams took delivery of yet more steam engines, among them some from Carlisle, the northern outpost of the London Midland Region, and when this valued source of scrap was finally exhausted at the end of 1968 some 220 locomotives (including four diesels) were crammed in the sidings around the West Pond site at Barry Docks. For many it was a rather undignified and depressing spectacle.

In all, Woodhams purchased nearly 300 locomotives, over 160 being former GWR types, the rest of LMS, SR and BR standard origin in approximately equal proportions, while only one ex-LNER example, Class B1 4-6-0 No 61264, found its way to this South Wales cemetery. They ranged from small shunting engines such as LMS 'Jinty' 0-6-0Ts to heavy freight and express passenger classes, two of the mighty GWR King class 4-6-0s and the solitary BR standard 8P Pacific No 71000 *Duke of Gloucester* being numbered among the latter.

Such a gathering of steam locomotives in public view could not fail to attract attention despite their worsening condition, and over the years thousands have wandered amidst their silent sullen ranks reminiscing of the days when these same engines trundled vast rakes of coal wagons through the South Wales valleys, or had charge of main line expresses out of Waterloo. Now the only sign of life was a mass of uncontrolled grasses, weeds and brambles, growing rapidly amidst the motion and cabs, while smokeboxes became the ideal nesting ground for a number of birds. The condition of these locomotives was not helped by the potently corrosive salty sea air, wafting in from the Bristol Channel, but with suitable candidates for preservation by way of direct purchase from BR almost depleted some began to have desires on restoring the best items of Woodhams stock to their former glories, although the pessimists doubted the wisdom of such schemes.

Heading the queue at Woodham's front door was the Midland 4F Preservation Society which enterprisingly handed over a cheque in exchange for Fowler 0-6-0 No 43924 and moved it to the Keighley & Worth Valley Railway in September 1968, where after overhaul it has become a valued member of the active stock. The sale required the concurrence of BR which had stipulated in the original agreement that engines must not be resold without its permission. The following year two

Despite their obvious sorry external condition four of these six engines photographed at Barry on 24 March 1976 have departed South Wales with high hopes of palmier days in the future. They are from left to right Nos 7200 – Quainton Railway Centre, 4270, 30506 – Mid-Hants Railway, 42765 – Keighley & Worth Valley Railway, 7927 Willington Hall, 34081 92 Squadron – Nene Valley Railway.

more engines departed for greener pastures, first in January Maunsell U class 2-6-0 No 31618 and later a second mogul, Churchward designed No 5322.

As will be seen from Table 2 these three engines became just the start of an exodus, since in recent years the de-stocking of the yard has been due almost exclusively to purchases by preservationists, only a handful having been broken up although on occasions there have been rumours that more wholesale scrapping was imminent. Over 30 ex-Barry engines have been steamed by their new owners, some even having worked over BR again,

Table 2

Run-down of Woodham's Steam Stock

	Locomotives moved away	Locomotives broken up	Total remaining at end of year
1968	1		216
1969	2		214
1970	8		206
1971	6		200
1972	10	1	189
1973	18	1	170
1974	19		151
1975	14		137
1976	8		129
1977	1		128
1978	11		117
1979	11		106
1980	8	2	96
1981	21		75
1982	3		72

while many more are well on the way to full restoration. A few have been retrieved purely as a source of spare parts or for static display, while Merchant Navy Pacific No 35029 *Ellerman Lines* was sectioned for exhibition at the National Railway Museum, York, (see page 71).

In recent times the saga of Woodhams has become a controversial issue, some campaigning that no more should be removed, arguing that resources could then be concentrated on those which have already been saved. These pleas have largely gone ignored and notwithstanding the worsening condition of those remaining at Barry, all having been there for well over a decade and a few double that time, the steady stream of departures for more secure homes has continued.

In an effort to obtain the best out of what remains for the railway preservation movement, either for restoration or spare parts, the Barry Rescue Project was founded in February 1981 with Robert Adley MP as chairman. Under its jurisdiction the engines were examined in detail and recommendations made regarding their futures.

Inevitably the day must dawn when the West Pond site will no longer serve as a locomotive graveyard, but for anyone who has ever strolled among these relics of a bygone age the mystic of the place will live on in the memory. Furthermore whatever the ultimate fate of the remaining decaying carcasses at Barry, whether it be restoration, cannibalisation or destruction, there is little doubt that the once humble but now revered scrapyard of Woodham Bros has earned itself a permanent and unique place in railway history, while lasting living memorials to the yard can be found throughout Britain. Indeed the praises for all that Woodhams have done will for a long time to come be thankfully echoed.

A panoramic view of Woodham's yard on 23 October 1982.

Line-up at the buffer-stops on 23 October 1982: from the left Stanier Class 5 4-6-0 No 44901, Stanier 8F 2-8-0 No 48173, BR standard 2-6-0 No 76084, and Churchward 42XX 2-8-0T No 4253. No 76084 has been sold to the Peak Railway, while the other three carry reserved notices.

7
THE NEW STEAM AGE MATURES

A British Railways branch line appears to be under threat of closure, and rumours to this effect are rife. The signs have looked ominous for some time, there is no freight traffic and the remaining passenger service can only be described as sparse. Few are surprised when closure notices are posted; there are the usual objections which force a public enquiry and delay BR's plans. As anticipated the protests are to no avail, the line has come to the end of its useful economic life and, in the view of the presiding inspector, there are adequate alternatives, so it must be lopped from the network. With little delay the date of closure is announced, railway enthusiasts and locals fill the last trains, the like of which has not been seen for years, before an eerie silence descends upon the route.

However even while BR still reluctantly operated the branch one or two far-seeing individuals realised the possibilities of the line as a tourist steam railway – with visions of splendour again. A meeting for interested parties was advertised and a preservation society formed. At first little could be done except raise funds and canvass opinion, but once BR's intention to vacate the stretch became official efforts were redoubled towards meeting the objective. Meanwhile BR agreed to leave the track for a limited period to see if negotiations with the preservation society could be satisfactorily concluded. At the same time the society applied for membership of the Association of Railway Preservation Societies – the umbrella organisation which co-ordinates the movement.

At last the society is able to hand over a cheque to BR for a short section and they become the proud owners of what amounts to nothing more than a length of track and a few repair-starved buildings. Hard graft (unpaid) then follows if the railway is to become functional, locomotives and rolling stock must be brought to the site and everything imaginable fully overhauled and renovated before submitting an application for a Light Railway Order, (LRO) the authority under which such lines normally operate. A hundred and one items need attention before steam operations can begin but eventually there comes the big day when the installations are to be diligently examined by an Inspecting Officer of the Railway Inspectorate. With bated breath his pronouncement is awaited: he makes one or two recommendations but yes, otherwise everything is in order, and services can commence once the LRO is issued.

Opening day sees the stations crowded, far more attending than on closing day a few years earlier. There is a joyous atmosphere about the place; after speeches by local dignitaries and society officials the inaugural train steams away in triumph. Another piece of railway has been reincarnated.

While the above is a very simple description of the fall and rise of a BR branch line, numerous groups, with various permutations, could hang their recent history around this theme. As outlined already, at the turn of the decade into the 1970s there were but half-a-dozen operational standard gauge steam lines, a nucleus of steam centres, and a few in the pipe-line, and with the supply of steam locomotives by direct purchase from BR then having run dry, the uninitiated could be forgiven for thinking that development in terms of new schemes would almost have been at an end. A glance at the map on page 61 will dispel this notion immediately, today's proliferation of live steam locations having come about steadily during the

Preservation of the old order at Bradford Industrial Museum. Built in 1922 by the Leeds-based Hudswell, Clarke & Co Ltd 0-4-0ST Nellie *proudly stands inside the transport gallery on 18 May 1982, an atmosphere much more purified than that at her previous home – Bradford Corporation's Esholt Sewage Works! On the right nameplates from other former steam locomotives adorn the walls of what was once a spinning mill.*

Another engine is added to the list of those saved for posterity. GWR Manor 4-6-0 No 7820 Dinmore Manor *arrives by low-loader at Bronwydd Arms, headquarters of the Gwili Railway, from Woodham Bros, Barry, on 23 September 1979. Ahead lie many hours of back-breaking work for Gwili Railway staff if the engine is to be restored to revenue earning condition.*

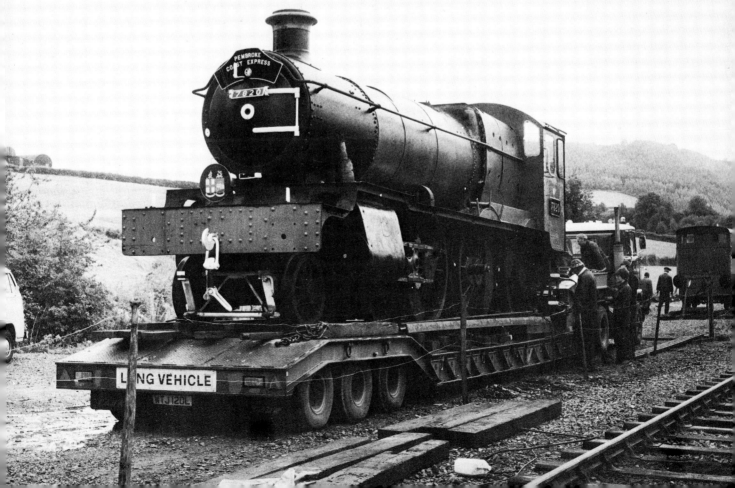

1970s and early 1980s. Even now others strive to join them, while some well-established railways plan to extend their sphere of operation.

Which comes first, the locomotive or the rails, is in some ways a chicken and egg question. What is certain is that the quest for suitable engines, both for established and new schemes, has continued unabated; but for their availability the number of active locations would be markedly less.

An earlier chapter narrated how the ex-BR locomotives saved during the 1960s have been supplemented through the gradual extraction of the fast decaying hulks incarcerated at Barry Docks by Woodham Bros, and dispersed throughout Britain. Some operators, moreover, could not resist casting envious glances in the direction of examples originally intended for static display only, and more than one sympathetic ear has resulted in a number being revitalised. Among them are Stroudley Terrier 0-6-0T No W8 *Freshwater* on the Isle of Wight Steam Railway, which before 1979 spent 13 years outside the Hayling Billy public house, Hayling Island, while the most notable in this category is Coronation Pacific No 46229 *Duchess of Hamilton* which was ensconced for over 10 years at Butlins holiday camp, Minehead, before its move to the National Railway Museum, York, in 1975.

In mentioning the National Railway Museum, opened in September 1975, its active participation in the railway preservation movement should be noted, for it has willingly allowed many of its engines to enjoy periods of activity at various locations, among them the treasured Great Northern Railway vintage pieces Atlantic No 990 *Henry Oakley* (Keighley & Worth Valley Railway) and Stirling Single No 1 (Great Central Railway). This, combined with their involvement in main line running and the extravaganzas which took place at Shildon and Rainhill in 1975 and 1980 respectively, has meant its influence has been felt far beyond the boundaries of Yorkshire. In return the NRM has benefited through engines, usually types not represented in the National collection, being readily loaned by the private railways for display in the magnificent setting created in the former York North steam shed.

A second large family of preserved locomotives are the former industrials, hardly a location being without at least one, while some concerns like the Foxfield and Tanfield Railways, and the Rutland Railway Museum, rely entirely upon them. Of the scores which survived into the 1970s, few have escaped close examination as to their suitability for

KEY TO MAP
1 Strathspey Railway
2 Lochty Private Railway
3 Scottish Railway Preservation Society
4 Bo'ness & Kinneil Railway
5 Prestongrange Mining Museum
6 Tanfield Railway
7 North of England Open Air Museum
8 Bowes Railway
9 North Yorkshire Moors Railway
10 National Railway Museum
11 Yorkshire Dales Railway
12 Keighley & Worth Valley Railway
13 Middleton Railway
14 Midland Railway Centre
15 Market Bosworth Light Railway
16 Great Central Railway
17 Rutland Railway Museum
18 Nene Valley Railway
19 North Norfolk Railway
20 Bressingham Steam Museum
21 Colne Valley Railway
22 Stour Valley Railway Preservation Society
23 Quainton Railway Centre.
24 Didcot Railway Centre
25 GWR Preservation Group
26 Kent & East Sussex Railway
27 Bluebell Railway
28 Hollycombe Woodland Railway
29 Mid-Hants Railway
30 Isle of Wight Steam Railway
31 Swanage Railway
32 Torbay & Dartmouth Railway
33 Dart Valley Railway
34 Bugle Steam Railway
35 West Somerset Railway
36 East Somerset Railway
37 Bristol Suburban Railway
38 Bristol Industrial Museum
39 Welsh Industrial & Maritime Museum
40 Caerphilly Railway Society
41 Swansea Maritime & Industrial Museum
42 Gwili Railway
43 Llangollen Railway
44 Cambrian Railway Society
45 Bulmer Railway Centre
46 Dean Forest Railway
47 Gloucestershire Warwickshire Railway
48 Severn Valley Railway
49 Birmingham Railway Museum
50 Chasewater Light Railway
51 Telford Horsehay Steam Trust
52 Foxfield Light Railway
53 North Staffordshire Railway
54 Dinting Railway Centre
55 Bury Transport Museum
56 Steamport Transport Museum
57 Steamtown Railway Museum
58 Lakeside & Haverthwaite Railway

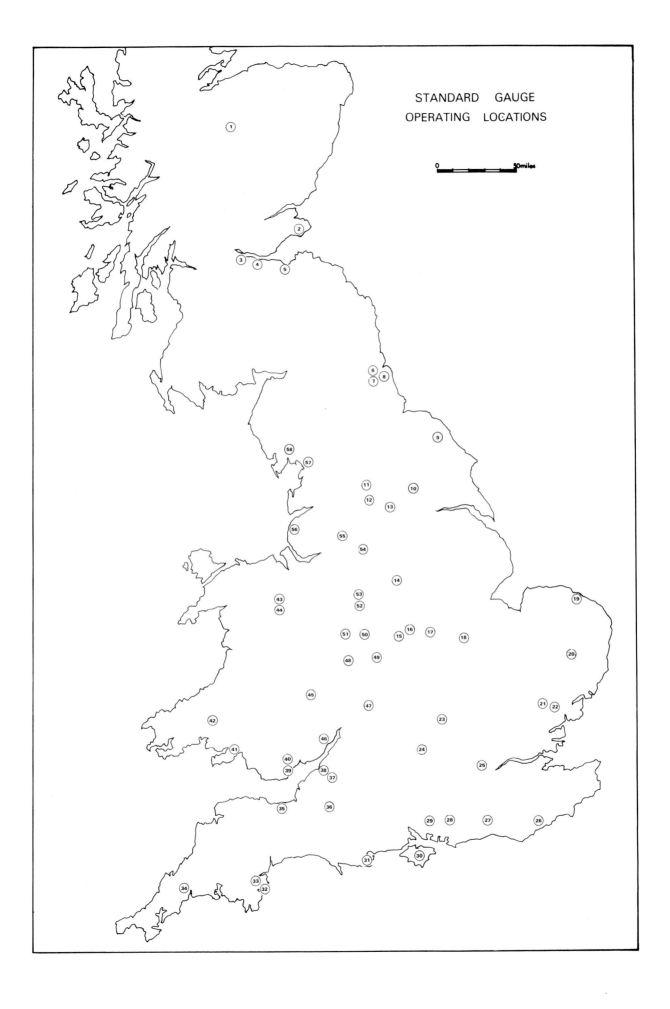

STANDARD   GAUGE
OPERATING   LOCATIONS

0   50miles

Imported from overseas. Ex-Ministry of Supply Austerity 2-8-0 No 1931, built by Vulcan Foundry in 1945, was repatriated in 1973 by the Keighley & Worth Valley Railway from the Swedish State Railways, which had it cocooned for 14 years as part of its reserve stock. No 1931 is the sole example of its type in Britain, although during the 1950s no fewer than 733 of its sisters roamed widely over BR, heavy freight haulage being their speciality (see page 14), but all were exterminated by the end of 1967. Here on 27 April 1974 No 1931 draws admiring glances from members of the fair sex while leaving Keighley for Oxenhope. Some minor modifications were necessary to the cab and boiler fittings before it could be used, as on arrival from Sweden the engine was a little taller than the British loading gauge.

a long term future. Many of those selected, having been groomed fit for the public to gaze on, are living an entirely different existence from that hitherto known at their oft-times cloistered environments. Although lacking the glamour of their main line sisters, their generally simple robust natures and lower initial cost of purchase, has made them attractive propositions compared with ex-BR types, although some of the latter which found their way into industrial service have also been plucked from this source. Some industrialists, realising the potential and historical value of their

charges, and perhaps not wishing to be responsible for the destruction of old and faithful servants, have magnanimously given, or placed them on permanent loan with suitable preservation centres. The National Coal Board has been very generous in this respect. As passenger locomotives, as any operator will testify, the industrials continue to give unstinting service.

In a bid to increase the stock of locomotives the intrepid have not been afraid to extend their search overseas, including countries behind the Iron Curtain, despite the enormous expense involved in shipping them to Britain. No fewer than 19 standard gauge types (along with numerous narrow gauge examples) have so far been imported, eight to be seen on the Nene Valley Railway, near Peterborough, which to overcome operational difficulties with those constructed to the limits of the Berne loading gauge of mainland Europe (roundly 1ft wider and higher than the British standard) made a number of alterations to clearances between track and platform faces and bridges. Welcomed home during the 1980s have been two engines built by Vulcan Foundry, Newton-le-Willows, Lancs, from the far flung corners of the world, a 5ft 6in

Attention to detail. The signals, ground-frame box, lamp-post, L&SWR warning sign, and the Fry's cocoa enamel advertisement provide an authentic as well as functional railway environment at Ropley, Mid-Hants Railway, as U class 2-6-0 No 31806 runs by with empty stock on 23 August 1981.

gauge 4-4-0 from Pakistan (VF No 3064 of 1911) for the Greater Manchester Museum of Science & Industry, and a 4ft 8½in gauge 4-8-4 from China (VF No 4674 of 1935) for the National Railway Museum. This source of power should remain available, albeit like much else of the steam empire on a diminishing scale, for some time to come.

A notable trend in recent years has been the construction of 12in to 1ft scale working replicas of a few historic locomotives. First for the Stockton & Darlington Railway 150th anniversary celebrations in 1975 came a facsimile *Locomotion*, followed by duplicates of *Rocket, Sans Pareil*, and *Novelty* for the Liverpool & Manchester 150 in 1980, while in 1981 the wraps were taken off a copy of Richard Trevithick's brainchild of 1804, to be seen at the Welsh Industrial & Maritime Museum, Cardiff. In an effort to make amends for the gap in respect of the broad gauge, under construction at the time of writing is a full size working reproduction of Gooch 4-2-2 *Iron Duke*, a similar engine to *Lord of the Isles*, which was originally saved by the GWR but through a lamentable volte-face destroyed in 1906. Further ambitious plans exist for recreating an LMS Patriot 4-6-0 at Southport, another sad omission from the preserved ranks.

In tracing the lineal descent of the standard gauge preserved locomotives, which total over 880, or when viewing a group in pristine condition, either in BR, constituent railway, or pre-grouping colours, it is hard to realise that not long ago many were mere rusting hulks. Thus it would be discourteous not to mention the vast amount of voluntary work expended on many engines to make them serviceable, most of the working locomotives having at some time been stripped to the frames, every nut, bolt and rivet examined, and then like a giant Meccano set painstakingly reassembled after repairs and restoration. In some cases this process has taken up to 10 years, while it is a sobering thought that others patiently await their turn for attention. Whatever future generations criticise about the 1960s and 1970s, never with justification will they be able to fault the resolve and efforts of those who contributed to the prolongation of steam as a living force. The derivation of the stud is shown in Table 3.

With the initial objectives achieved, many then

Table 3: Standard Gauge Preserved Locomotives

Former Main Line types[1]

Great Western Railway	105	
Southern Railway	68	
London Midland & Scottish Railway	88	
London & North Eastern Railway	41	
BR Standard	33	335

Former Industrial types[2,3]

Andrew Barclay, Sons & Co Ltd	99	
Avonside Engine Co Ltd	22	
Hudswell, Clarke & Co Ltd	45	
Hunslet Engine Co Ltd	60	
R. & W. Hawthorn, Leslie & Co Ltd	28	
Peckett & Sons Ltd	63	
Robert Stephenson & Hawthorns Ltd	53	
W. G. Bagnall Ltd	32	
Other Builders	127	529

Imported from Overseas	19
TOTAL	883

Notes:
[1] Totals include locomotives built for the constituent railways.
[2] Does not include locomotives built by private locomotive builders for the main line companies.
[3] Main line locomotives which subsequently had spells in industrial service are included above.

began to consider the long term needs of their stock – covered accommodation and maintenance needs being high on the list of priorities. For the latter some concerns have developed well-equipped workshops able to deal with the majority, if not all, maintenance and repair requirements, as at Sheffield Park, Bridgnorth and Buckfastleigh. A few accept outside contracts which bring in additional revenue. Some locomotives have been entrusted to others for attention, including BREL main works; sometimes this is confined to work of a specialist nature – like boiler repairs, while many owners have co-operated in locating spare parts and in financing the manufacture of patterns for the fabrication of new ones. Work on the locomotives is a continuing process and can be observed at virtually every location on any week-end – and at a few during the week as well where full-time staff are employed.

The prime attractions at any steam centre are the

Ladies are welcome as volunteers too – and not just for serving the refreshments! At Steamtown Railway Museum, Carnforth, Mrs Erica Arneil cleans out the smokebox of A3 Pacific No 4472 Flying Scotsman *on 17 June 1979. Mrs Arneil presently works in a voluntary capacity as assistant to Bill Harvey, locomotive engineer and consultant to railway preservationists, and author of* A Manual of Steam Locomotive Restoration and Preservation *published by David & Charles.*

Shades of 1825. Built in 1975 for the Stockton & Darlington Railway 150 celebrations, the replica Locomotion *is viewed on one of the demonstration lines at the North of England Open Air Museum, Beamish, Co Durham, together with a couple of chaldron wagons on 4 September 1977. In 1975 the original* Locomotion, *after spending 85 years on a pedestal on Darlington Bank Top station, was moved, along with S&DR No 25* Derwent *of 1845, to the new Darlington Railway Museum, which utilises part of North Road station on the old S&DR route.*

locomotives, some even stealing the show at museums which deal with many aspects of life and where railways form just a small section, as at the North of England Open Air Museum, Beamish, Co Durham, and some industrial museums. However today's private railways are catering principally for people who are visiting them merely for pleasure (although a minority use them purely as a means of transport) and if members of a discerning public are to be enticed back again, and to feel able to recommend a visit to friends, then innumerable items require the same detailed attention as the locomotives.

As mentioned earlier the scene greeting the new occupants of any stretch of track has generally been one of desolation, with vegetation rampant, any remaining facilities being nothing more than basic. By earnest endeavour the permanent-way and signalling systems have had to be brought up to an acceptably high standard, new sidings laid, buildings renovated, and to create the right atmosphere, as well as for functional purposes, such structures as water-towers, turntables, signalboxes etc, procured from other redundant BR sites. The provision of adequate parking spaces, souvenir shops, toilet and refreshment facilities, picnic sites, areas where out of service engines can be worked on in safety, attractions such as model and miniature railways, and small museums devoted to railwayana of past generations, all require careful thought. The image of the staff is equally important, and on some railways distinctive uniforms are worn by paid employees and volunteers alike when on certain duties.

That the new railways overall have been very successful is borne out by the thousands who return each year to sample their wares, some having annual passenger figures in the six figure category. To meet the demand intensive timetables are operated at peak periods, sometimes far in excess of that ever run by BR; the Torbay & Dartmouth Railway and the Keighley & Worth Valley Railway have had to install extra crossing loops in order to be able to operate a more frequent service than was originally possible when the lines were taken over from BR. Indeed the larger concerns have of necessity had to become very pro-

fessional in outlook, organising themselves into departments to cover various responsibilities – locomotives, carriage and wagon, permanent-way, signalling etc. In every respect they seek the ideals of the great railway companies of old.

High on the agenda at every meeting is the mundane topic of finance, but to neglect it, as with any business, courts disaster. As the movement has grown, and with vast sums of capital required to operate even a small depot, good husbandry has become more and more imperative, the advice of accountants (and at times solicitors) being equally as important as that of the engineers and boilersmiths.

Over the years finance has been obtained from every conceivable source, not least that received by direct subscription from members and patrons. Some like the infant Gloucestershire Warwickshire

The line through Boat of Garten used to be part of the Highland Railway main line to Inverness via Forres, but today all BR trains bound for the Highland capital are concentrated on the route over Slochd summit. The Strathspey Railway has brought life back to the southern tip of the abandoned section, and on 12 June 1982 Hunslet Austerity 0-6-0ST No 60 (works No 3686 built 1948) leaves Boat of Garten with the 15.20 for Aviemore, Scotland's premier ski resort nestling between the Monadhliath and Cairngorm mountain ranges. The Hunslet Austerities are the most numerous standard gauge preserved class, this particular example coming from the NCB Dawdon Colliery, Seaham, Co Durham.

For many years the everyday activities of the Scottish Railway Preservation Society were restricted to its depot at Falkirk, but after a number of abortive attempts a suitable branch line was acquired at Bo'ness, on the southern bank of the Firth of Forth, a short length opening on 27 June 1981. The aim is eventually to run as far as Bo'ness Junction on the Edinburgh–Glasgow main line. On 19 June 1982 the stone buildings of Bo'ness form a backdrop for Andrew Barclay 0-4-0ST No 1937 of 1928, emblazoned with the name of a previous owner the South of Scotland Electricity Board Clydesmill Power Station, as it passes the disused docks with an LNER non-corridor brake composite coach. The SRPS retains its Falkirk site.

Hoping one day to follow in the footsteps of the other concerns depicted on these two pages are the Ayrshire Railway Preservation Group. Looking forlorn, but nevertheless cared for, along with other stock gathered at the closed Minnivey Colliery, near Dalmellington, on 22 April 1981, is ex-NCB 0-4-0ST No 19, Andrew Barclay No 1614 of 1918. A photograph of this engine during its NCB days, appears on page 34.

Railway project have floated public limited companies, grants have been obtained from various bodies, including the relevant tourist boards, while many individual projects have been sponsored by industry and commerce. In the case of the West Somerset Railway the track bed was acquired by Somerset County Council in 1973 and then leased back.

Much welcome additional revenue has been obtained by the provision of facilities for feature films, documentaries and television commercials. For this purpose some lines and locomotives have been disguised almost beyond recognition in order to masquerade as Victorian or overseas sets, while Gresley Pacific *Flying Scotsman* – painted appropriately on opposite sides – doubled as Nos 4474 *Victor Wild* and 4480 *Enterprise* in the film *Agatha*. A by-product is the free exposure the railways receive from the various media, the Keighley & Worth Valley Railway for example seeing a marked

With Marley Hill Colliery in the background ex-CEGB Hudswell Clarke 0-4-0ST No 1672 Irwell *drops down gradient on the Tanfield Railway, four miles south west of Gateshead, with a four-wheeled Tanfield-built coach on 4 September 1977. Today there are over a score of industrial locomotives at this location.*

Earlier the same day octogenarian NER Worsdell Class J21 0-6-0 No 876 (BR No 65033), together with a 1910-built NER 20 ton coal wagon, paces up and down a short demonstration line at the North of England Open Air Museum, Beamish, Co Durham. Railway items account for but a small portion of this museum which portrays a varied cross section of life as it was in the latter part of the Victorian era. Many buildings – industrial and domestic, including Rowley station from the closed NER Consett to Crook line, were carefully dismantled at their original sites and then meticulously re-erected at Beamish. The museum provides a fascinating look back in time.

The North Eastern Locomotive Preservation Group celebrated its 10th anniversary with a gala day on the North Yorkshire Moors Railway on 31 October 1976. Here Raven Class T2 0-8-0 No 2238 is signalled into Levisham with a Grosmont–Pickering train, a somewhat different life-style to ferrying coal and iron ore around Durham on which it was occupied during its BR days as No 63395.

upsurge in interest following release of *The Railway Children* in late 1970, having been filmed mainly in and around the Worth Valley.

To supplement the voluntary labour and that of any full-time employees, valued assistance has come by way of Government Manpower Services Commission Job Creation Schemes, Army units who have undertaken civil engineering projects as training exercises, and apprentices from one or two firms who have been allowed time to work on steam locomotives. They have brought two-fold benefits: the railways have seen the early completion of certain plans, while the personnel involved have gained experience which otherwise may have been denied them.

However welcome these sources of one-off income and practical help, they must be looked

Driver's eye view from the footplate of Class K1 2-6-0 No 2005 at Pickering on the North Yorkshire Moors line on 19 September 1982.

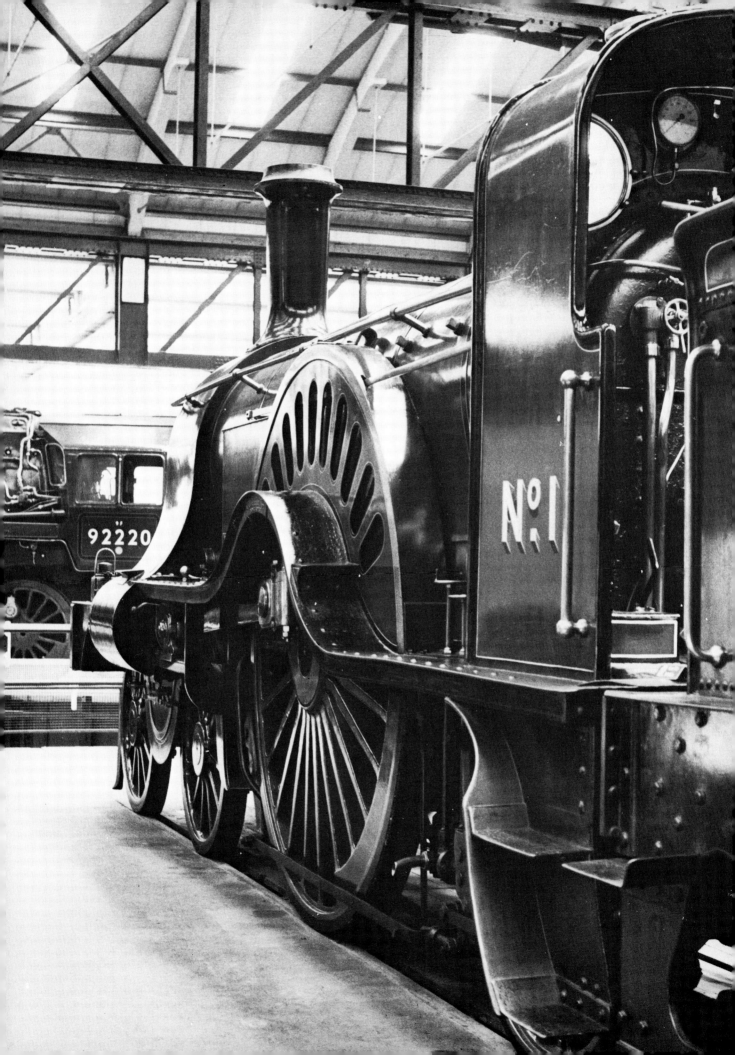

See how it works. Rebuilt Bulleid Pacific No 35029 Ellerman Lines *was sectioned to give this explicit view of the inside of a modern steam locomotive, which has proved of immense educational value at the National Railway Museum.* Ellerman Lines *was removed from the rusting lines at Barry Docks in 1974 specifically for this purpose, the driver's side being restored to its 1960s condition when it regularly headed Weymouth and Exeter expresses out of Waterloo.*

Without doubt the National Railway Museum, York, established in September 1975, with its vast range of exhibits, large and small, ancient and modern, has been an unqualified success; over 10 million people entered its hallowed precincts within seven years of opening. On 21 May 1977 Patrick Stirling's magnificent 8ft single, No 1, built for the Great Northern Railway in 1870, looks towards one of the twin turntables, here occupied by Evening Star. *No 1 had a short spell of activity in 1981-2 on the Great Central Railway, Loughborough.*

upon as the icing-on-the-cake, and for their day to day needs and long term viability the steam lines must do everything possible to maximise their regular income. With this in mind and to sustain the interest of members, such attractions as vintage transport days, schools specials, wine-and-dine trains, along with enthusiasts' days with as many engines in steam as possible have become regular features, as have visits by Father Christmas to many locations during December with a present for every child travelling on the trains.

From those first tentative steps taken during the 1950s and 1960s the new steam era has without doubt come of age and firmly established itself. Let us then take a tour around Britain and view what is just a small portion of the work undertaken by the latter day railway navvies and engineers, worthy successors indeed to Brunel, Hackworth, the Stephensons, Gresley, Stanier etc. As we shall see, the traditions of these forefathers rests in very good hands.

Lucky to be alive. This NER Worsdell Class H 0-4-0T No 1310 was withdrawn by the LNER in 1931, but instead of facing the big hammer was fortuitously purchased by the Pelaw Main Collieries and subsequently bought for preservation from the NCB in 1965. Here No 1310 propels a light load from Tunstall Road Halt to Middleton Park Gates on the Middleton Railway, Leeds, on 25 October 1975.

Slough Estates Ltd No 5, Hudswell Clarke 0-6-0ST No 1709, built 1939, moves sedately away from Bow Bridge loop (near the junction with the BR freight only truncated Grassington branch) towards Embsay station on the Yorkshire Dales Railway, near Skipton, on 18 April 1982. A galaxy of ex-industrial tanks are grouped at Embsay, and plans exist for extending their sphere of operation further along this ex-Midland Railway line in the direction of Bolton Abbey.

Lancashire & Yorkshire Railway centenarian 0-6-0ST No 752 pilots Robert Stephenson & Hawthorns 0-6-0ST No 63, a mere youngster of 28 years, near Ingrow with the Keighley & Worth Valley 16.27 Keighley–Oxenhope on 12 April 1982.

29 August 1981, only one week after the reintroduction of passenger services at Butterley by the Midland Railway Centre on the ex-Midland Railway Ambergate to Pye Bridge line, and already the fishermen must realise that they no longer have Butterley reservoir to themselves! As LMS 4F 0-6-0 No 4027 crosses the causeway in the direction of Butterley station one may speculate if this young man, here busily winding in, will become hooked on steam in the near future!

From behind the Iron Curtain. USA-built Class S160 2-8-0
No 5820, affectionately known as 'Big Jim', attacks the gradient
out of Keighley with a K&WVR train for Oxenhope during
March 1979. 'Big Jim' arrived from the Polish State Railways
in November 1977 and is representative of the hundreds of
S160s which worked on British railways during the latter
stages of the second world war; following D-Day all were shipped
to Europe. While perhaps not the prettiest engine 'Big Jim' has
proved a very useful asset on the K&WVR, handling with ease
the heaviest trains.

On 29 August 1981 at Swanwick Junction Midland Railway senior citizen 2-4-0 No 158A of 1866, on loan to the Midland Railway Centre from the National Railway Museum, keeps company with Stanier Princess Pacific No 6203 Princess Margaret Rose, which spent 12 years as an attraction at Butlins holiday camp, Pwllheli, before its arrival at the Midland Railway Centre in 1975. Both locomotives are at present inoperable.

On a glorious summer afternoon Hawthorn Leslie 0-6-0ST No 3931 built 1938 sets out from Shackerstone for Market Bosworth along the Market Bosworth Light Railway, a route once jointly owned by the LNW and Midland railways, on 8 August 1982. The MBLR started services during Spring 1978 and it expects to extend them by 1985 to Shenton, adjacent to the Battle of Bosworth site being developed by Leicestershire County Council where King Richard III met his death in 1485, whereupon King Henry VII took the throne of England. At present the railway relies solely on former industrial types, although No 3931 has since forsaken Shackerstone for the Swanage Railway.

The Loughborough to Rothley section of the erstwhile Great Central main line has taken on a lease of new life in recent years, steam trains again treading the route taken by The South Yorkshireman and The Master Cutler on their journeys from Yorkshire to Marylebone. On 13 August 1978 the driver of Gresley N2 0-6-2T No 4744 (BR No 69523) brings a Loughborough bound train to a halt at Quorn & Woodhouse on the 'new' Great Central Railway. No 4744, built in 1921, was fitted with condensing gear for use when hauling Kings Cross area suburban trains through the tunnels to Moorgate; the large pipes for carrying the exhaust steam back to the side tanks are clearly visible. Few British classes were so equipped.

On the Nene Valley Railway Swedish State Railways Class S1 2-6-4T No 1928 eases over the level crossing at Wansford, at the end of its five mile journey from Orton Mere, near Peterborough, on 24 August 1980. The NVR opened for business on 1 June 1977.

Two more locomotives from Sweden outside NVR's Wansford shed on 27 August 1982: left Class B 4-6-0 No 1697 and Class S 2-6-2T No 1178. Who could have envisaged in the days when BR owned this ex-LNWR route that locomotives such as these would one day roam freely through Cambridgeshire? Engines from Denmark, France, Germany, and Italy, as well as from Britain, have also made the Nene Valley their home.

The first of the BR standard locomotives, Pacific No 70000 Britannia, arrives at Orton Mere from Wansford on 4 May 1981; the 30A shedplate is that of its first home, Stratford, to which it was allocated when new in 1951 for working East Anglian expresses out of Liverpool Street. Notice the extra width and height of the continental vehicle behind Britannia compared to the rest of the formation; the NVR adopted the Berne loading gauge to accommodate the more amply proportioned engines and stock from overseas. The extra effort is reflected in the number of continental railway film sequences shot on the Nene Valley without the expense of transporting a film crew to mainland Europe.

With an admiring audience Colne Valley Railway Hunslet
Austerity 0-6-0ST No 3790 of 1952 Castle Hedingham *pushes
its train out of Castle Hedingham station, near Halstead,
Essex, on 30 August 1982. The last BR freight trains ran along
here in 1965, and when the CVR took over the area in 1973
they were greeted by nothing more than a weed infested track-
bed, since when a remarkable transformation has taken place,
including the erection of numerous buildings, much material
being brought from other local redundant station sites. The
footbridge came from Stowmarket.*

Built by the Glasgow-based North British Locomotive Co Ltd
as its No 24564, in 1939, 0-6-0T Coventry No 1 *heads a
bizarre assortment of period coaching stock at the* Quainton
Railway Centre, *near Aylesbury, on 28 September 1980. Over
30 steam engines are housed here, ranging from small industrial
tanks and a working fireless locomotive to GWR King class
4-6-0 No 6024* King Edward I. *Slicing through the centre of the
site (the line next to* Coventry No 1) *is the BR freight only line
from Aylesbury to Claydon LNE Junction on the
Bletchley–Oxford route.*

Part of the National collection, Stanier taper-boilered three-
cylinder 2-6-4T No 2500 *gives footplate rides at Bressingham
Steam Museum, near Diss, on 2 September 1982. Introduced in
1934, the 37 members of this class are best remembered for their
work on the London, Tilbury & Southend lines out of
Fenchurch Street, hence the Southend destination board carried
by No 2500.*

Super-power in the form of 4-6-0s No 7808 Cookham Manor *and No 5051* Drysllwyn Castle *at the Great Western Society Didcot Railway Centre on 26 September 1981.* Drysllwyn Castle *was the original name bestowed on No 5051 when new in 1936, but the following year it was renamed* Earl Bathurst, *both names having been carried at various times since restoration. The automatic mail pick-up and collection apparatus, seen on the right, is demonstrated on special occasions.*

Adams Class 0415 4-4-2T No 488 belies its 89 years with the Bluebell Railway 14.45 Sheffield Park – Horsted Keynes on 15 September 1974. Its BR days were spent mainly on the Axminster–Lyme Regis branch as No 30583.

BR closed the last remaining section of the immortal Lt Col Holman F Stephens' Kent & East Sussex Railway (the first line built under the 1896 Light Railways Act) in 1961, whereupon a preservation society was formed to save the portion from Tenterden to Robertsbridge. However years of frustration, and not a little anguish, lay ahead, as the railway became embroiled in a long costly legal battle with the Ministry of Transport, whose worry regarding the effect on road traffic at seven level crossings succeeded in delaying the ideals of the new régime until February 1974, although by then it had been decided to preserve only the 10-mile section between Tenterden and Bodiam. Many lesser mortals would have given up the ghost long before!

Here proving that the efforts expended were all worthwhile USA 0-6-0T No 22 Maunsell (BR No 30065) ascends Tenterden bank in fine style on 25 May 1981. This is one of 14 former US Army Transportation Corps 0-6-0Ts bought by the Southern Railway in 1946 for use in Southampton Docks; four grace the 1980s. (Geoff Silcock)

1954-built BR standard Class 4 4-6-0 No 75027 arrives at
Horsted Keynes with the 10.35 from Sheffield Park on 15
September 1974. Although not withdrawn until the end of steam
on BR in August 1968, No 75027 has now spent more time as
a preservation piece!

Not far from its 1898 Brighton birthplace Class E4 0-6-2T
No 473 Birch Grove still haunts the Sussex Weald along the
tracks of its former masters – the London, Brighton & South
Coast Railway, here scurrying down Freshfield bank bound for
Sheffield Park on the Bluebell Railway in July 1970.

With the safety valves lifting, a rejuvenated SR U class 2-6-0, No 1618 of 1928, makes light work of its Bluebell Railway load at Holywell on 9 July 1977. In 1969 this became the second engine to be removed from Woodham Bros scrapyard at Barry.

A most impressive sight to behold – rebuilt Bulleid West Country class Pacific No 34016 Bodmin with a Mid-Hants Railway Alresford to Ropley train on 23 August 1981. Probably few passengers are aware that Bodmin spent over seven years wasting away at Barry Docks after BR brought its life to an end, or so they thought, in June 1964, a tribute indeed to all who helped to restore the engine to its present fine condition. As for how Bodmin looked before rebuilding in 1958, see photographs of sister engine City of Wells on pages 11 and 97.

Not content with simply shuttling the three miles between Alresford and Ropley, as it has since April 1977, the Mid-Hants is making strenuous efforts to extend its activities a further seven miles 'over the Alps' to Alton, where it will share the BR station. Here track panels are loaded at Alresford during 1982 before laying on the Ropley–Medstead section; Maunsell N class 2-6-0 No 31874 Brian Fisk is in charge of the permanent-way train. (Courtesy Mid-Hants Railway)

Isle of Wight Steam Railway Hawthorn Leslie 0-4-0ST No 37 Vectis (works No 3135 built 1915) in a sylvan setting at Havenstreet, with a train returning from Wootton on 24 August 1981. More traditional Island power in the shape of Stroudley Class A1X 0-6-0Ts No W8 Freshwater and No W11 Newport, and Adams O2 0-4-4T No W24 Calbourne, can also be seen here.

On the Isle of Purbeck, Swanage Railway Hunslet 0-6-0ST *Cunarder* (No 1690 of 1931) gently draws a couple of coaches into the canopied Swanage station on 16 August 1981. Faintly chalked on the smokebox door is the No 30131, that of a Drummond M7 0-4-4T – a class which frequented Swanage in the days when its trains connected with those on the Bournemouth–Weymouth line at Wareham.

King and Castle 4-6-0s once used these tracks of the Torbay & Dartmouth Railway on the last stage of their journeys from Paddington to Kingswear with the Torbay Express, tracks which passed into the ownership of the Dart Valley Railway in Autumn 1972, and to date the only preserved line not to see any interruption in services following BR's withdrawal of trains. Unfortunately plans to run an all-year-round commuter service between Paignton and Kingswear quickly had to be abandoned, but during the Summer months copper-capped products of Swindon regularly entice the holidaymakers away from the beaches for a closer look and a trip to Kingswear.

In sight of Torbay's golden sands and clear waters, somewhat deserted on this cool afternoon, yet another Barry escapee GWR 5205 class 2-8-0T No 5239 makes a full-throated assault on the 1 in 67 from Goodrington Sands Halt with the 16.00 Paignton (Queens Park)–Kingswear on 3 May 1982. While Great Western purists may bemoan the fact, this former heavy freight locomotive, power classification 8F, was appropriately named *Goliath* in June 1979. Many previously unnamed engines have been so honoured by their new owners, which helps the general public to identify them more readily.

Below: West Somerset Railway Bagnall 0-6-0ST Victor *No 2996 of 1951 meanders through the Quantock Hills with a light-weight train from Bishops Lydeard to Minehead on 15 March 1981.* Victor *is one of two modern industrials domiciled on the WSR which formerly earned their keep at British Leyland's Longbridge plant, Birmingham. The WSR originally intended running over the whole of the Taunton-Minehead branch, but it proved impossible and WSR management has concentrated resources on the 20-mile Bishops Lydeard-Minehead section; even so it is the longest preserved line in Britain.*

Bottom: Locally built by the Bristol-based Avonside Engine Co Ltd, 0-6-0ST Edwin Hulse *(works No 1798 of 1918) propels one coach and a brakevan up grade from Bitton station, Bristol Suburban Railway, on 6 May 1979. Before the 1923 grouping this was a Midland Railway spur from the Birmingham–Bristol main line at Mangotsfield to Bath Green Park, whereby through running from the North to the South Coast was then possible via the Somerset & Dorset Joint Railway.*

A re-creation of what started it all back in 1803/4 – a full size working model of Trevithick's 1804 Pen-y-darren locomotive at the Welsh Industrial & Maritime Museum, Cardiff, on 23 October 1982. Note the flangeless wheels and the 4ft 4in gauge L-shaped plateway track. The engine is usually steamed once a month.

Left: At the Dart Valley Railway Buckfastleigh line in May 1982 the sole surviving Hawksworth 16XX class 0-6-0PT No 1638 leaves Buckfastleigh with the 13.30 to Totnes Riverside. This was Hawksworth's last design for the GWR, although it was 1949 before the pioneer example made its debut outside Swindon Works, while it was not until 1955 that the last of the 70 built took to the rails – seven years after nationalisation.

On 23 September 1979 Gwili Railway Peckett 0-4-0ST Merlin/Myrddin, No 1967 built 1939, leaves a layer of uncondensed steam over the Dyfed countryside while heading north to Cwmdwyfran from Bronwydd Arms, an ex-GWR thoroughfare which at one time stretched from Carmarthen to Aberystwyth.

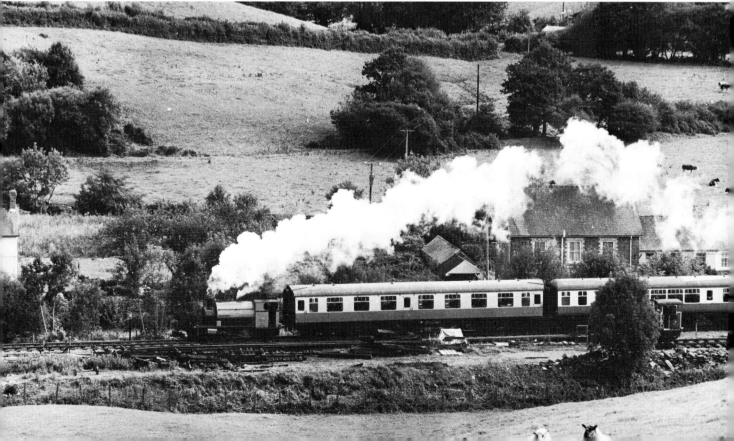

Valley of song and steam. Famed as the home of the
International Music Eisteddfod, Llangollen, in the Dee Valley,
has also been roused by the syncopating exhausts of active steam
locomotives since July 1981, and near the limit of present
operations at Pentrefelin, Kitson No 5459, 0-6-0ST
Burtonwood Brewer *built 1932 returns its train to Llangollen
on 11 July 1982. The Llangollen Railway hopes to extend
gradually towards Corwen, a route which once linked Ruabon
with Barmouth and Blaenau Ffestiniog.*

Heavy rain throughout the afternoon of 24 October 1982 could
not dampen the ardour of those visiting the Dean Forest
Railway Norchard Steam Centre – and no wonder with sights
such as this to behold; former Barry internee GWR Prairie
tank No 5541, with two GWR chocolate and cream auto-
trailers, overtakes Hunslet Austerity 0-6-0ST No 3806 of 1953
G B Keeling, *busy shunting a demonstration freight train. The
long-term aim of the DFR is to operate the Lydney–Parkend
branch.*

On 25 September 1982 at the Gloucestershire Warwickshire Railway headquarters at Toddington, Hall class 4-6-0 No 5952 Cogan Hall has high hopes of steaming through the Cotswolds, along parts of the GWR Stratford-upon-Avon to Cheltenham line, in days to come. Giant strides towards this goal have been made in recent times, some track having been laid towards Winchcombe, while four ex-BR locomotives, including Cogan Hall, have been acquired from Woodham Bros and a few industrials from other sources.

Stanier 8F 2-8-0 No 8233 takes on coal in Bridgnorth yard, Severn Valley Railway, on 20 May 1973. No 8233 has had a very chequered career; built by the North British Locomotive Co in 1940 for the War Department it was sent to Iran in 1941, moved to the Suez Canal zone in 1944, and returned to England in 1952. After spending the next five years on the Longmoor Military Railway it was bought by BR in 1957, and as No 48773 steamed until the last in August 1968, although in the early 1960s it was withdrawn and subsequently reprieved on two separate occasions. The engine is owned by the Stanier 8F Locomotive Society.

Trains first ran through the luxuriant Severn Valley from Shrewsbury to Hartlebury on 1 February 1862, the Great Western taking control from the pioneer Severn Valley Railway company during the 1870s, the former being responsible for the link from Bewdley to Kidderminster. Through services were maintained until 1963, whereupon the line gradually fell victim to BR's retrenchment policy. A preservation society was founded on 6 July 1965 but, despite strenuous efforts, it was 23 May 1970 before the new proprietors were allowed to run trains, at first between Bridgnorth and Hampton Loade. Adorned with the old BR lion and wheel emblem on the tender, Ivatt LMS taper-boilered 2-6-0 No 46521 rides high over Bewdley South viaduct working from Bewdley to Foley Park during an SVR enthusiasts' week-end on 12 September 1976. It is only on special occasions that trains ply this section, although through excursions to and from BR by way of Kidderminster pass this way. SVR hoped to purchase this section in 1984.

Severn Valley Railway Collett Manor 4-6-0 No 7819 Hinton Manor banks a train headed by Ivatt 2-6-0 No 6443 up to Foley Park Tunnel on 15 April 1978. When SVR trains traverse this length it is necessary to have an engine at each end as there are no run-round facilities at Foley Park.

No 7819, along with No 46521 depicted above, owe their existence to the fact that when BR decided they were surplus to requirements during the 1960s, fate decreed that they were among the batches purchased by Woodham Bros and towed to Barry Docks. No. 46521 reached the SVR in Spring 1971, with No 80079, and No 7819 arrived together with Nos 4141, 5164 and 4930 Hagley Hall in January 1973. Yes, the name of Dai Woodham will long be remembered.

Hawthorn Leslie 0-4-0ST Asbestos (No 2780 of 1909) picks its way through Chasewater Pleasure Park, Brownhills, on the Chasewater Light Railway, with a former BR Gloucester dmu trailer coach No E56301 in tow on 18 August 1974. Asbestos was donated by Turners Asbestos Cement Co Ltd, Trafford Park, Manchester.

Despite the lettering, Andrew Barclay 0-4-0ST No 2352, built 1954, on static display in a children's playground at Rocester, near Uttoxeter, on 31 August 1981, came from no further afield than the CEGB Goldington Power Station, Bedford. A number of small engines can be found similarly marooned in other play areas around Britain.

Not a Martian invasion, but sandblasting in progress on the firebox of LMS 0-6-0 4F No 4422 at Cheddleton station, North Staffordshire Railway, on 31 August 1981; the engine arrived from Woodham Bros, Barry, in April 1977. The picture epitomises the enormous efforts made to restore many engines to operable condition.

The well-established Dinting Railway Centre, near Glossop, has the use of the old Great Central Dinting engine shed, occupied on 28 October 1978 by partly restored LNWR 0-6-2T No 1054 (BR No 58926) built 1888, on loan from the National Trust, Penrhyn Castle Museum, near Bangor. No 1054, a Webb Coal Tank, later appeared at the Rocket 150 celebrations. The majority of the stock at Dinting is housed in a large purpose-built exhibition hall.

The pleasing sight of Lancashire & Yorkshire Railway stalwart Aspinall Class 27 0-6-0 No 1300 in action at Steamtown Railway Museum, Carnforth, on 31 May 1982. After withdrawal by BR as No 52322 in 1960, the engine spent many years as a static exhibit in the yard of Leonard Fairclough Ltd, Adlington, near Chorley, after a cosmetic restoration at its native Horwich.

On 28 July 1982 ex-Preston Docks shunter 0-6-0ST Princess
starts out of the ex-Furness Railway Haverthwaite station, now
part of the Lakeside & Haverthwaite Railway, with the 10.40
for Lakeside, where it will connect with a Sealink vessel on
Lake Windermere. Princess was built at W G Bagnall's Castle
Engine Works, Stafford, in 1942 as its No 2682. BR withdrew
passenger services over the branch in 1965, the L&HR opening
for business in May 1973.

8
MAIN LINE STEAM FROM 1971

At the turn of the decade into the 1970s the steam preservation movement had become established and the future looked rosy. However there was a growing lobby of opinion which had no wish to see steam confined simply to pottering along short sections of privately-owned track with an imposed maximum speed of 25mph, as laid down by the light railway orders governing operation. Indeed the burning desire of a great many people was to be able once more to savour the engines with the reins unfurled, and a heavy load behind, out on the main lines they had served so faithfully in the past.

At this time the British Rail hierarchy were adamant – there was definitely no place for steam on its modernised system. They stubbornly argued that it was just not possible to accommodate steam amidst its diesel and electric traffic and, moreover, the necessary servicing facilities no longer existed. But despite many rebuffs the protagonists remained undeterred and their tireless efforts were justly rewarded during late Summer 1971 when BR suddenly had a change of heart, probably much sooner than even the most optimistic had dared hope. This was due in no small way to the determination of Peter Prior, managing director of H. P. Bulmer Ltd, the Hereford cider makers, who had in their care in full working order, GWR King class 4-6-0 No 6000 *King George V*, an agreement being reached with BR for a short series of trial runs in the Autumn with No 6000 to assess the feasibility of steam hauled specials.

Steam returns to BR – a most heart-warming occasion. Pictured during its history-making journey from Hereford to Tyseley on 2 October 1971 GWR King class 4-6-0 No 6000 King George V gets a clear road at Wolvercot Junction, just north of Oxford; five Bulmer's Pullman cars are immediately behind the engine. (Terry Flinders)

The inaugural run on 2 October 1971 was to mark the start of a new epoch for steam as *King George V* with regal splendour headed south from Hereford to Newport with a rake of five Pullman cars and two BR coaches, and then continued via the Severn Tunnel and Oxford to Tyseley. Later that week the train visited Kensington and Swindon before returning to Hereford. Throughout the journey the lineside was thronged with excited photographers and well-wishers, the like of which had not been seen for over three years; for many the period had seemed like an eternity!

With the tour a complete success and the case of the pro main line steam faction vindicated, BR readily sanctioned for 1972 a limited number of

GWR Castle class 4-6-0 No 7029 Clun Castle *descends Hatton bank on 10 June 1972. Even driving rain and bleak lighting conditions could not quell the high spirits of the hundreds who flocked to the Birmingham-Didcot route to catch a glimpse of this the first steam special following* King George V's *barrier breaking epic the previous October.*

steam hauled excursions along the undermentioned secondary routes:—

Birmingham Moor Street–Didcot	77	miles
Shrewsbury–Newport	94	,,
York–Scarborough	42	,,
Newcastle–Carlisle	60	,,
Carnforth–Barrow	28	,,
Total	301	,,

They were chosen because of their close proximity to the bases of suitable engines, with turning facilities available at each end in the form of triangles, although in the event the two Stanier Class 5s used on the trips to Barrow from Carnforth had to return tender first. Servicing was the responsibility of the operators. Again there were no real problems.

The ice was now well and truly broken, BR countenancing steam for 1973 and furthermore increasing the available mileage to over 760 by the

An important aspect of main line operations is the servicing and maintenance of the locomotives in which steam centres such as Steamtown, Carnforth, Bulmer's Railway Centre, Hereford, and Didcot Railway Centre, fulfil a vital role. Being prepared for duty on the SLOA sponsored Golden Arrow Pullman (see page 11) at a snow-shrouded Steamtown, Carnforth, on 12 December 1981 is Bulleid unrebuilt light Pacific No 34092 City of Wells. Steamtown now possesses the only modern type coaling-plant in Britain (left) whereby tenders can automatically be loaded from an overhead storage bunker, a hand-down from the days when this location served as a BR steam shed. The plant is put to good use by the many engines which venture onto BR from here.

Engines are also stabled and serviced at a few BR depots, sometimes sharing the facilities with diesel locomotives, as at Northwich shed on 18 May 1980, Stanier Black 5 4-6-0 No 5000 keeping company with Class 25 No 25281 and Class 24 No 24081. The previous month No 5000 had powered a special with No 80079 between Hereford and Manchester (see page 110). No 24081 has since been preserved at Steamport Transport Museum, Southport.

Photo-stops are a regular feature of steam specials, these being well-stewarded, while steps are taken to ensure that no other trains are in the vicinity. Rostered to a Welsh Marches Express from Shrewsbury to Hereford on 7 March 1981, Stanier Princess class Pacific No 6201 Princess Elizabeth *pauses at Church Stretton while travellers take a close look at her. This journey is described in the prologue.*

addition of another eight sections of track. The steam season, largely confined to the months of April, May, June, September and October was equally as successful with 28 trains run, this setting the pattern for the next few years. As well as trains along the recognised steam routes (which have seen a few additions and deletions over the years) BR has occasionally allowed 'one-offs', as for the Maidenhead–Marlow branch centenary on 15 July 1973 for which the diminutive 0-4-2T No 1450 travelled up from Buckfastleigh, Devon, and in 1975 along the East Coast main line between York and Newcastle in connection with the Stockton & Darlington Railway 150th anniversary celebrations. These gestures are much appreciated.

In the main the trains have been composed of BR standard Mark 1 coaches, sometimes with the addition of a privately-owned support vehicle of vintage origin. However an enterprising innovation took place on 19 October 1974 when a complete set of restored GWR carriages, resplendent in chocolate and cream, from the Great Western Society depot,

Didcot, were marshalled behind 4-6-0s No 7808 *Cookham Manor* and No 6998 *Burton Agnes Hall* for an outing to Stratford and Tyseley, and over the next seven years the stock made numerous journeys on BR, as did a similar set preserved by the Severn Valley Railway. Sadly their wanderings ended, at least for the time being, after a final flurry on 26 January 1980, because of the high costs involved in complying with BR conditions of use. While at large they certainly evoked vivid memories of the GWR.

The most significant developments since the return of steam came in 1978. First was the welcome announcement that steam would henceforth be permitted over the scenic Settle & Carlisle route, advocated by many for some time, although refused point blank for the line's centenary in 1976; thus at Easter Gresley V2 2-6-2 No 4771 *Green Arrow* became the first steam locomotive to venture this way to Carlisle since the BR farewell special of 11 August 1968. A year later the revival over the Long Drag was to be cruelly cut short by the Penmanshiel Tunnel disaster on the East Coast main line, north of Berwick, and the subsequent need to use the S&C as a diversionary route. Happily once the Edinburgh–Newcastle line was reopened steam again returned to the S&C and with a vengeance.

Secondly in 1978 came the equally welcome news

Working from Hellifield to Carlisle with the Cumbrian Mountain Pullman Stanier Coronation class Pacific No 46229 Duchess of Hamilton *rolls into Appleby on 24 October 1981. At one time this class had charge of principal expresses on the West Coast main line out of Euston and were then, as today, firm favourites with enthusiasts. In its original streamlined form* Duchess of Hamilton, *masquerading as No 6220* Coronation, *together with a set of Coronation Scot coaches, visited America in 1939, identities being exchanged again with the real No 6220 on its return. When not on main line duty No 46229 is housed at the National Railway Museum, York.*

'Jinty' 0-6-0T No 7298 glints in the evening sunshine as it leads stock, including Black 5 4-6-0 No 44806 Magpie *and two National Railway Museum owned Class 502 electric multiple-units, out of Southport station back to Steamport Transport Museum, at the end of a joint BR/Steamport exhibition on 12 September 1982. Isolated movements such as this occur from time to time.*

Not the climb to Masbury summit in the Mendip hills on the vanished Somerset & Dorset Joint Railway, where 7F 2-8-0 No 13809 spent all its career from 1925 until withdrawn by BR in 1964, but the approach to Gisburn on the equally rural nominally freight only Blackburn to Hellifield line. No 13809 is heading the Wyvern Express from Sheffield to Carnforth via Manchester on 31 October 1981. The engine spent a dismal 10 years at Woodham Bros scrapyard, Barry, before its timely rescue in 1975, its home now being the Midland Railway Centre, Butterley.

Thompson B1 class 4-6-0 No 1306 Mayflower strides away from Guiseley Junction, Shipley, while en route from Carnforth to York with a private charter on 12 June 1977. The line to Ilkley can just be seen to the right of Mayflower. This engine is not at present available for main line duty, being based on the non-BR connected Great Central Railway, Loughborough.

that during the peak holiday season BR was to sponsor steam excursions aimed at the family market rather than enthusiasts in particular, somewhat ironically having to hire suitable locomotives from the preservationists. The London Midland Region formulated plans for a train starting from Blackpool to Sellafield on the Cumbrian Coast every Tuesday from 27 June until the end of August, with steam from and to Carnforth, while during the same period the Eastern organised two trains each Sunday following a circular path from York through Leeds and Harrogate back to York. The response was overwhelming, and in an effort to satisfy demand the London Midland, after the second week, doubled the number of seats available by running a repeat train on Wednesdays.

Since then every summer has witnessed a series of BR steam hauled trains based on Carnforth and, apart from 1980, York, with variable itineraries and a few operating between the two centres in order to exchange locomotives for short periods to provide a variety of motive power on each side of the Pennines. In some years the schedules have appeared in the BR timetable. In 1983 SLOA (see below) organised the mid-week trains on the London Midland.

To co-ordinate all these charter specials an umbrella organisation, the Steam Locomotive Owners Association (SLOA) was formed in 1975, and latterly has organised and marketed the tours enabling the accrued profits to be retained by the locomotive owners towards future maintenance requirements. BR also benefit in that it only needs negotiate with one body. The Scottish Steam Railtours Group fulfils a similar role north of the border.

With the Settle & Carlisle once more available to steam from the beginning of 1980 SLOA enterprisingly sponsored a series of six Cumbrian Mountain Expresses during the early part of the year, one engine being used from Carnforth to Skipton and a second to Carlisle, with the return leg over Shap electrically worked, while the next CME traversed the circuit in the opposite direction in order to bring the S&C engine south. So popular were these trains that SLOA quickly increased their number to 12 and then BR followed by including the route in its summer programme every Thursday, all utilising steam south from Carlisle over the S&C, the locomotive having gone north on the previous Tuesday's Cumbrian Coast Express to Sellafield and then light to Carlisle.

Gresley D49 4-4-0 No 246 Morayshire *(BR No 62712) makes a fine sight in Collessie Den near Lindores Loch, not far from Newburgh, on the single track Ladybank to Perth line on Easter Sunday 19 April 1981.*

Gresley A4 streamlined Pacific No 4498 Sir Nigel Gresley sprints through Lostock Junction, near Bolton, with The Lancastrian, working from Shrewsbury to Carnforth via Chester and Manchester on 23 May 1981. The line going out of the picture on the right leads to Wigan. Who could have foreseen when this illustrious class held sway on the East Coast main line from Kings Cross, that in the 1970s and 1980s one would find regular employment in the North West!

Serving as a reminder of a once everyday sight along the LSWR West of England main line, rebuilt Bulleid Merchant Navy Pacific No 35028 Clan Line speeds through Grateley, between Basingstoke and Salisbury, on 27 April 1974 – destination Westbury. To the disappointment of many people no BR lines in the South of England have seen steam since 1975.

(Terry Flinders)

Table 4

Engines which have hauled passenger trains on BR since 1971.

Date built	Original owner	Wheel arrangement	Number	Name
1838	L&MR	0-4-2		*Lion*
1892	LNWR	2-4-0	790	*Hardwicke*
1935	GWR	0-4-2T	1450	
1924	GWR	4-6-0	4079	*Pendennis Castle*
1929	GWR	4-6-0	4930	*Hagley Hall*
1936	GWR	4-6-0	5051	*Earl Bathurst/Drysllwyn Castle*
1931	GWR	4-6-0	5900	*Hinderton Hall*
1927	GWR	4-6-0	6000	*King George V*
1931	GWR	2-6-2T	6106	
1944	GWR	4-6-0	6960	*Raveningham Hall*
1949	BR★	4-6-0	6998	*Burton Agnes Hall*
1950	BR★	4-6-0	7029	*Clun Castle*
1930	GWR	0-6-0PT	7752	
1938	GWR	4-6-0	7808	*Cookham Manor*
1939	GWR	4-6-0	7812	*Erlestoke Manor*
1925	SR	4-6-0	777	*Sir Lamiel*
1936	SR	4-6-0	841	*Greene King*
1926	SR	4-6-0	850	*Lord Nelson*
1949	BR★	4-6-2	34092	*City of Wells*
1948	BR★	4-6-2	35028	*Clan Line*
1902	MR	4-4-0	1000	
1951	BR★	2-6-0	43106	
1947	LMS	4-6-0	4767	*George Stephenson*
1945	LMS	4-6-0	44871	*Sovereign*
1945	LMS	4-6-0	44932	
1934	LMS	4-6-0	5000	
1934	LMS	4-6-0	5025	
1937	LMS	4-6-0	5305	
1937	LMS	4-6-0	5407	
1935	LMS	4-6-0	5596	*Bahamas*
1936	LMS	4-6-0	5690	*Leander*
1927	LMS	4-6-0	6115	*Scots Guardsman*
1933	LMS	4-6-2	6201	*Princess Elizabeth*
1938	LMS	4-6-2	46229	*Duchess of Hamilton*
1925	S&DJR	2-8-0	13809	
1937	LNER	4-6-2	4498	*Sir Nigel Gresley*
1937	LNER	4-6-2	60009	*Union of South Africa*
1937	LNER	4-6-2	60019	*Bittern*
1923	LNER	4-6-2	4472	*Flying Scotsman*
1936	LNER	2-6-2	4771	*Green Arrow*
1948	BR★	4-6-0	1306	*Mayflower*
1949	BR★	2-6-0	2005	
1928	LNER	4-4-0	246	*Morayshire*
1891	NBR	0-6-0	673	*Maude*
1954	BR	2-6-4T	80079	
1959	BR	2-10-0	92203	*Black Prince*
1960	BR	2-10-0	92220	*Evening Star*

★Engines built by BR to constituent railway company designs.

In recent times repeat charter packages such as the Cumbrian Mountain Express and others like the Welsh Marches between Shrewsbury and Newport or Hereford and Chester have become the norm, these trains following similar schedules each time, the tickets also covering travel by ordinary service trains to and from suitable pick-up points from stations throughout Britain. While photographic stops have been observed since the early days, it has now become standard practice to include in the Cumbrian Mountain and Welsh Marches tours the added attraction of one or two photographic run-pasts, which allow participants a chance to watch a little action from the lineside. The location of these is varied and they are very much enjoyed.

A cause of growing concern of late has been the reduction in BR steam-heated vacuum-braked stock. Thus with an eye on the future SLOA purchased in early 1980 eight Metro-Cammell Pullman cars and two first- and second-class Mark 1 BCKs from BR, all dual steam/electric heated and capable of 100mph running, especially important when en route to and from the steam sections behind air braked, electrically-heated diesel or electric locomotives so that the line speed can be upheld. These new acquisitions made their debut on 2 May 1981 with Stanier Class 5 4-6-0 No 5407 at the helm south over the Settle & Carlisle. The stock, along with more recent additions, is maintained by BR, and from 1982 its standard BR blue and grey livery has gradually given way to the traditional Pullman umber and cream.

The accompanying table details the 47 locomotives employed to date to haul passengers on BR since the steam embargo was annulled, ranging from small tank locomotives such as Nos 1450 and

Far from its ancestral home on 27 December 1980, Maunsell 4-6-0 No 850 Lord Nelson *runs across the long viaduct spanning the Kent estuary, just west of Arnside station on the former Furness Railway route, with a Santa Special bound for Sellafield; these trips have become regular Christmas fare in recent years.*

7752 on the odd occasion, to the regulars and firm favourites with the general public and enthusiasts alike, as for example *Flying Scotsman* and *Sir Nigel Gresley*. Deserving special mention is the veteran Liverpool & Manchester Railway *Lion* of 1838 which made the short journey from Eccles to Manchester Liverpool Road on 14 September 1980 (see page 126). Not listed are the many other locomotives which have utilised BR tracks in connection with stock movements and while travelling to open days or exhibitions, as for the jamborees at Shildon and Rainhill in 1975 and 1980 respectively. Today old company or regional loyalties almost count for nothing; the only criteria for use, apart from compliance with the loading gauge, is that the engines satisfy the demanding safety standards laid down by BR. Thus it has become common for former Southern Railway engines to roam the North of England or Gresley designs the West Cumbrian coast.

A glance at the map on page 106 will quickly reveal the extent of the BR network traversed by steam passenger trains from 1971. Excluded are those routes used only by light engines or for stock movements.

Today steam is again an integral part of the BR scene and at times on the lines out of Carnforth and York its passage is almost an everyday occurrence during certain parts of the year. As well as hauling the SLOA and BR summer specials they also see service on schools trains, private charters and for

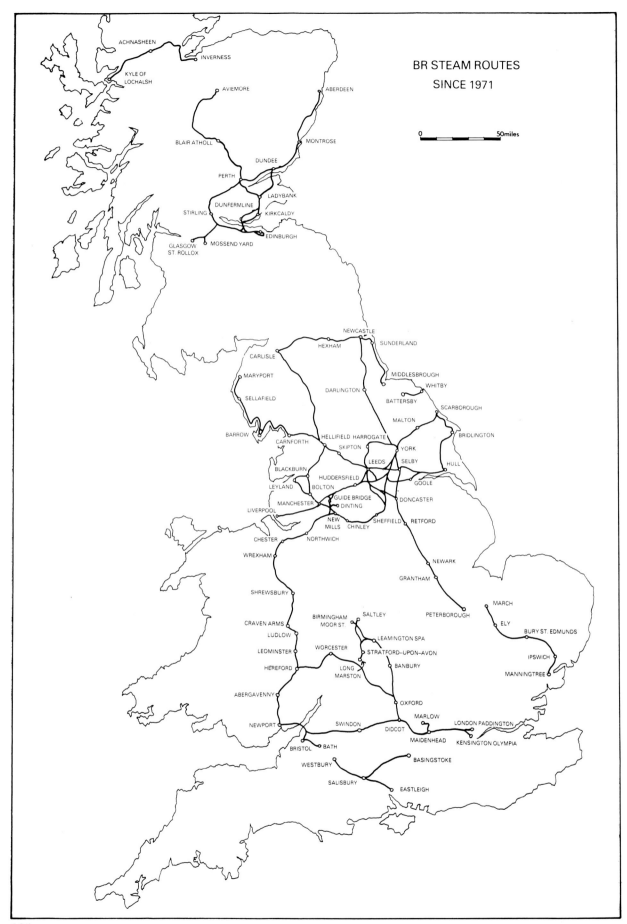

BR STEAM ROUTES
SINCE 1971

0   50miles

ACHNASHEEN
INVERNESS
KYLE OF LOCHALSH
AVIEMORE
ABERDEEN
BLAIR ATHOLL
MONTROSE
DUNDEE
PERTH
LADYBANK
STIRLING
DUNFERMLINE
KIRKCALDY
GLASGOW ST. ROLLOX
MOSSEND YARD
EDINBURGH

NEWCASTLE
HEXHAM
SUNDERLAND
CARLISLE
MIDDLESBROUGH
MARYPORT
DARLINGTON
WHITBY
SELLAFIELD
BATTERSBY
SCARBOROUGH
MALTON
BARROW
BRIDLINGTON
CARNFORTH
HELLIFIELD HARROGATE
SKIPTON
YORK
LEEDS
SELBY
HULL
BLACKBURN
HUDDERSFIELD
GOOLE
LEYLAND
BOLTON
MANCHESTER
GUIDE BRIDGE
DONCASTER
LIVERPOOL
DINTING
NEW MILLS
SHEFFIELD
RETFORD
CHINLEY
CHESTER
NORTHWICH
WREXHAM
NEWARK
GRANTHAM
SHREWSBURY
MARCH
CRAVEN ARMS
BIRMINGHAM MOOR ST.
SALTLEY
PETERBOROUGH
ELY
LUDLOW
LEAMINGTON SPA
BURY ST. EDMUNDS
LEOMINSTER
WORCESTER
STRATFORD–UPON–AVON
IPSWICH
HEREFORD
LONG MARSTON
BANBURY
MANNINGTREE
ABERGAVENNY
OXFORD
NEWPORT
MARLOW
SWINDON
LONDON PADDINGTON
DIDCOT
MAIDENHEAD
KENSINGTON OLYMPIA
BRISTOL
BATH
WESTBURY
BASINGSTOKE
SALISBURY
EASTLEIGH

Maunsell King Arthur 4-6-0 No 777 Sir Lamiel *is swung round on the turntable at Scarborough after working the Yorkshire Pullman to the East Coast seaside resort on 18 September 1982. The closure of Filey Holiday Camp station and its attendant triangle in 1977 brought to an end steam visits to Scarborough (for a second time!) until the former Gateshead turntable was installed in spring 1981. Scarborough Council contributed towards the costs.*

The wider dimensions of many ex-GWR locomotives compared with others of the four group companies, allowed by a more generous loading gauge, has meant that generally they are confined to their former habitats.

The end of the long gruelling climb from Shrewsbury is in sight for Castle class 4-6-0 No 4079 Pendennis Castle *nearing Church Stretton on 6 April 1974. The engine was exported to Australia in 1977 and is now operated on special passenger workings by its present owner Hamersley Iron Pty Ltd.*

advertising purposes, while on three separate occasions steam has gone to the aid of failed diesel locomotives on freight trains – a real triumph for these once unwanted giants!

But what of the future? Much obviously depends on BR, the continuing availability of suitable lines; one or two like the Settle & Carlisle do not look too secure – while financial considerations, as always, will play their part, especially when boilers or other expensive items require renewal. BR employees with steam experience are diminishing but encouragingly a number of volunteers, and the crews, are undergoing training in the art of firing; moreover BR has already sanctioned main line running until 1990. If all continues to go well is there any reason why the date will not be extended yet again? Many will hope so, for what finer spectacle is there to stir the emotions than a large steam locomotive out on the main line with a long string of coaches trailing behind?

Immaculately groomed, and with a Great Western style train reporting number Z78 in position, King class 4-6-0 No 6000 King George V has no trouble with the 1 in 112 rising gradient at Little Stretton while travelling from Hereford to Chester with a rake of restored chocolate and cream GWR coaches from the Severn Valley Railway on 23 April 1977. The bell over the front buffer beam commemorates No 6000's trip when new to the Baltimore & Ohio Railroad centenary celebrations in 1927. This section of track was originally part of the Shrewsbury & Hereford Joint line, owned by the GWR and LNWR.

108

With the return Shakespeare-Don from Didcot to Birmingham *Castle* 4-6-0 *No 7029* Clun Castle *runs parallel with the Oxford canal approaching Aynho Junction on 17 April 1976.*

Perhaps surprisingly since they were the last steam engines built for BR, only three standard types have to date been used on the main line since relaxation of the steam ban.

The last steam locomotive to be built for BR, class 9F 2-10-0 No 92220 Evening Star, is outlined by its exhaust while pulling away from Garsdale across the 12-arch Dandry Mire viaduct, northbound for Carlisle on 13 May 1978.

With the safety valves gently lifting, railway and wildlife artist
David Shepherd's 9F 2-10-0 No 92203 Black Prince *was
photographed in the western outskirts of Worcester heading for
Hereford on 19 May 1973.*

The third member of the trio, Class 4 2-6-4T No 80079 tucked
in behind Stanier Class 5 4-6-0 No 5000 departs from Chester
for Manchester on 19 April 1980.

As a tribute to the late John Scholes, who was Curator of Historical Relics at the British Transport Commission Clapham Museum of Transport, a special train, The Curator, ran from London to York, with steam from Guide Bridge, on 26 May 1979. In charge of The Curator, Midland Compound 4-4-0 No 1000 and LNER V2 2-6-2 No 4771 Green Arrow burst through Marple station; both engines are part of the National collection and in their early preserved days were under the watchful eye of John Scholes.

Sadly the eminent railway photographer, the Rt Rev Eric Treacy, former Bishop of Wakefield, died while recording No 92220 Evening Star at Appleby station on 13 May 1978. The railway fraternity paid its respects by organising two special steam-hauled trains over the Settle & Carlisle – a line the Bishop loved dearly – on 30 September 1978, and during the afternoon Evening Star formed a backcloth for a memorial service in the former Appleby goods yard. Here a section of the congregation are gathered just before the service, after which Mrs Treacy unveiled a suitably engraved commemorative plaque on Appleby station. Clan Line and Flying Scotsman were also used on the specials.

Steam returned to London on 1 March 1979 to celebrate the 125th anniversary of Paddington station. With city high rise blocks towering in the background and the A40(M) motorway on the left, GWR King class 4-6-0 No 6000 King George V is surrounded by its successors, four IC125s and a Class 31, at the approach to Westbourne Park with a Paddington-Didcot special. Surfacing on the right after passing under the main lines is London Transport's Metropolitan line.

In acknowledgement of the fact that for 150 years troops had been carried by rail, appropriately SR 4-6-0 No 850 Lord Nelson was selected to head a special from Liverpool Lime Street to York on 11 March 1981, here awaiting marching orders at Liverpool. As the train made its way through the tunnels to Edge Hill it was regaled by a military band.

Food prepared and cooked on a train was first served in a Pullman car in 1879 on the Great Northern Railway Kings Cross to Leeds route, and 100 years later a representative selection of restaurant cars, ancient and modern, toured BR, steam locomotives providing the power over certain approved routes; Gresley V2 2-6-2 No 4771 Green Arrow roars up to Micklefield Junction with the gourmet special, travelling from York to York via Leeds and Harrogate on 29 September 1979. In the right foreground is the line to Selby and Hull.

Two Stanier Class 5 4-6-0s in harness. Nos 45407 and 44932 leave York for Carnforth on 29 April 1978.

Surrounded by GWR lower quadrant signals Ivatt 2-6-0
No 43106 and Collett 4-6-0 No 7812 Erlestoke Manor *pass
Wrexham station with a Welsh Marches Pullman bound for
Chester on 6 June 1982. Both are Severn Valley Railway
engines.*

LMS Class 5 No 5025 plods determinedly towards
Druimuachdar summit (1,484ft above sea level and now the
highest point on BR) near Dalnaspidal with a Perth-Aviemore
excursion on 12 June 1982. It is over 17 miles from Blair
Atholl to the summit, much of it at 1 in 70, one of the toughest
climbs asked of steam locomotives today. Home for No 5025 is
the Strathspey Railway at Aviemore, where it was turned before
returning to Perth later in the day.

K1 2-6-0 No 2005 pilots Class 5 4-6-0 No 4767 George
Stephenson *away from a signal check at Haydon Bridge,
between Newcastle and Carlisle, on 24 October 1981. The train
had started at Middlesbrough and continued over the Settle &
Carlisle, by which means the engines were transferred from the
North Yorkshire Moors Railway to Steamtown, Carnforth.*

With Birmingham city centre in the background, Modified Hall 4-6-0 No 6998 *Burton Agnes Hall* and Hall 4-6-0 No 5900 *Hinderton Hall* approach Small Heath, returning to their Didcot home with the Great Western Society restored GWR coaches on 15 May 1976.

On one of its all too rare appearances on BR, Dinting Railway Centre Royal Scot class No 6115 *Scots Guardsman* blasts away from New Mills with *The Yorkshire Venturer*, which it worked from Guide Bridge to York and back on 11 November 1978. The Royal Scots were a 1927 Fowler design for the LMS, but from 1943 were rebuilt to Stanier plans with taper boilers, new cylinders and double chimneys, as seen here.

The solitary LMS Class 5 4-6-0 fitted with Stephenson's link motion, No 4767 George Stephenson, brings life to the otherwise desolate landscape of upper Ribblesdale while crossing Ribblehead viaduct with the northbound Cumbrian Mountain Express on 21 February 1981.

Veteran of the 1895 Railway Race to Aberdeen LNWR 2-4-0 Precedent class No 790 Hardwicke still going strong at Meathop with a Carnforth to Grange-over-Sands shuttle on 23 May 1976.

Some of the umber and cream SLOA-owned Pullman stock normally used on SLOA trips.

9
LIVERPOOL & MANCHESTER 150

In October 1829 the well-documented Rainhill trials took place at which Stephenson's *Rocket* triumphed. Eleven months later came the official opening of the Liverpool & Manchester Railway and thereafter little stood in the way of iron and later steel railway spreading its tentacles not only in the land of its birth but throughout the world.

British Rail rightly decided that the 150th anniversary of these important milestones in railway history should, like the centenary in 1930, be celebrated in style. Arrangements were made for the main event, Rocket 150, to be staged during the 1980 Spring Bank Holiday week-end, when on three consecutive days a grand cavalcade of rolling stock passed along the actual length of the 1829 trials between Lea Green and Rainhill. In all 31 steam locomotives of widely varying ages and parentage, including three purpose-built replicas of the leading contenders at the original trials, were paraded before hundreds of enraptured spectators, followed by an equally sparkling array of more modern motive power. As they slowly but proudly 'steamed' between the specially erected grandstands at a modest 5mph the locomotives were complemented by a mixed assortment of passenger and freight stock, and together they represented the full spectrum of railway development over the past 150 years.

Many ancilliary exhibitions, both large and small, also took place in the North West that summer, along with a series of main line steam specials which traversed most of the original route between Liverpool and Manchester. Overall they formed a fitting tribute, not only to the Stephensons and the other railway pioneers, but to all who have followed in their footsteps and influenced in any way the emerging high speed, and in places, intensely operated railway we know today.

It is appropriate to recall here some of the events of 1980, for they embraced many aspects of the railway preservation movement. Worth noting, too, is the fact that without the ready co-operation of some of the privately-owned steam lines and the National Railway Museum, not forgetting the initiative and determination of those who saved and restored the

Last minute attention to detail as GWR Castle class 4-6-0 No 5051 Drysllwyn Castle *(left) receives a final brush-down, while (right) the Pines Express headboard is attached to S&DJR 2-8-0 No 13809 before the final cavalcade.*

exhibits in the first instance, Rocket 150 and the rest could never have even been contemplated, at least not on anything like such a grandiose scale. British Rail must be eternally grateful.

Now we may speculate what form the bicentenary celebrations will take in 2030, both in terms of preservation and railway development!

On 21 May 1980 Midland Railway 4-2-2 No 673 in its ninetieth year, LMS 0-6-0 No 4027 and Somerset & Dorset Joint Railway 2-8-0 No 13809 were hauled from the Midland Railway Centre, Butterley, via the Hope Valley route to Bold Colliery, St Helens. This rare sight in Chinley station yard became available when the diesel pilot was requisitioned to rescue a preceding failed diesel multiple-unit on the line ahead, Class 40 No 40022 on the left being similarly stranded.

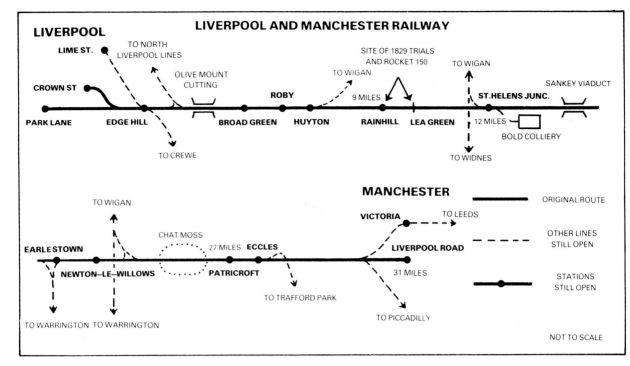

LIVERPOOL AND MANCHESTER RAILWAY

LIVERPOOL

LIME ST.
TO NORTH LIVERPOOL LINES
OLIVE MOUNT CUTTING
SITE OF 1829 TRIALS AND ROCKET 150
TO WIGAN
CROWN ST
ROBY
9 MILES
TO WIGAN
SANKEY VIADUCT
ST. HELENS JUNC.
PARK LANE
EDGE HILL
BROAD GREEN
HUYTON
RAINHILL
LEA GREEN
12 MILES
TO CREWE
BOLD COLLIERY
TO WIDNES

MANCHESTER

ORIGINAL ROUTE

VICTORIA
TO LEEDS

TO WIGAN

OTHER LINES STILL OPEN

CHAT MOSS
27 MILES
ECCLES
LIVERPOOL ROAD
EARLESTOWN
NEWTON—LE—WILLOWS
PATRICROFT
31 MILES
STATIONS STILL OPEN
TO TRAFFORD PARK
TO WARRINGTON
TO WARRINGTON
TO PICCADILLY

NOT TO SCALE

Positively sparkling, former Longmoor Military Railway 2-10-0 No 600 Gordon, *now based on the Severn Valley Railway, moves slowly forward ready to take its turn in the second of the processions past the grandstands on 25 May 1980.*

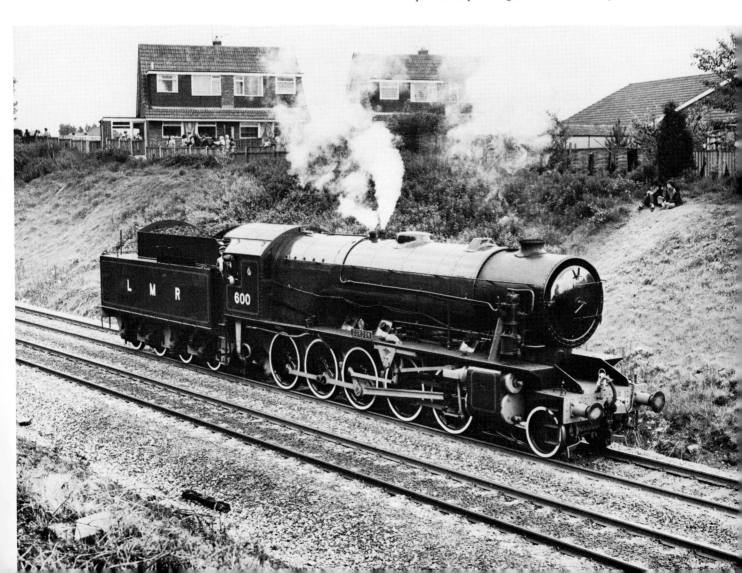

The star of the show. The replica of Stephenson's Rocket *captures the attention and imagination of the spectators while passing in front of the stands drawing a facsimile Liverpool & Manchester Railway open second class carriage, an added touch of authenticity provided by the crew dressed in period costume.*

Following Rocket, *a copy of a second contender in the 1829 trials, Timothy Hackworth's* Sans Pareil *pulls an equally realistic replica L&MR first class coach, the design closely resembling the early nineteenth century road coaches. In the rear* Lion *is ready to lend a helping hand in case of need.*

Like the original Novelty *built by Braithwaite & Ericsson of London, the 1980 version also failed to live up to expectations and had to spend its time sitting in this well-truck, being exhibited in the cavalcades behind Class 5 4-6-0 No 5000, along with the only narrow gauge locomotive to take part,* The Earl *from the 2ft 6in gauge Welshpool & Llanfair Railway. While outwardly the three replica engines are virtually identical reproductions of the originals, many modern methods of construction had to be used in order to comply with present day safety standards.*

A Royal Train. Bearing the correct headlamp code LNWR 2-4-0 No 790 Hardwicke *of 1892 vintage passes the official starting point in charge of three LNWR coaches once used to convey members of the Royal family, and normally to be seen at the National Railway Museum.*

All eyes are focused on Severn Valley Railway based GWR Collett 0-6-0 No 3205 travelling back to Bold Colliery, St Helens, with a passenger brake, the out and home workings of the cavalcades giving customers a second opportunity to view the exhibits.

For a short time during May 1980 the NCB Bold Colliery, St Helens, adjacent to the former L&MR and just east of St Helens Junction station, became home for the locomotives attending Rocket 150; with at times some 30 engines in steam the atmosphere became very much akin to that which once prevailed at BR sheds. They were an emotive few days.

A sight to quicken the heart-beat of any steam enthusiast with seven engines prepared ready to move out of the yard, the two nearest the camera are (left) SR Schools class 4-4-0 No 925 Cheltenham and BR standard 2-6-4T No 80079, with LMS 'Jinty' 0-6-0T No 7298 between on 26 May 1980.

On eight consecutive Sundays from 22 June 1980 BR sponsored a Steam One-Five-O special over the 30 miles from Manchester Victoria to Edge Hill (Liverpool) and back, with stops on the outward run at Earlestown, and Rainhill on the return, enabling passengers to visit the small exhibitions at these stations. There was also time available to view the Edge Hill Railway Trust exhibition while the engine was turned and serviced at Liverpool. With the last of the specials on 10 August 1980 Stanier Jubilee class 4-6-0 No 5690 Leander gathers speed over the nine-arch Sankey viaduct after the Earlestown stop.

After scrambling on to the wall in the right foreground, three small boys are captivated by the sight and sound of Stanier Princess class Pacific No 6201 Princess Elizabeth heading past Eccles signalbox with the second of the Steam One-Five-O excursions on 29 June.

To mark the sesquicentenary of the opening of Manchester Liverpool Road station, BR operated a 150th Anniversary Special from Liverpool Lime Street to Manchester Victoria on Sunday 14 September 1980. The dignitaries, among them Sir Peter Parker, chairman of British Rail and Home Secretary William Whitelaw, travelled outwards from Manchester by a diesel multiple-unit which halted at Parkside while wreaths were laid on the memorial to William Huskisson MP, who died after being hit there by Rocket on 15 September 1830. 150 years later Stanier Coronation Pacific No 46229 Duchess of Hamilton confidently lifts the special away from Liverpool at the approach to Olive Mount cutting, a spectacle to which the father of railways himself George Stephenson, would no doubt have warmed! On the left the Liverpool skyline is dominated by the Roman Catholic Metropolitan cathedral.

At Eccles the official party transferred to a short train of replica L&MR coaches for the remainder of the journey to Liverpool Road station in the care of Todd, Kitson & Laird 0-4-2 Lion, itself built for the L&MR in 1838, while Duchess of Hamilton went on ahead. Here with a brass band playing harmoniously in the front carriage and those occupying the second suitably attired, Lion steams merrily through the Salford suburbs, the modern colour light signal abruptly reminding us that this is 1980 and not the 1840s!

It is fitting to conclude this section with a view of the heroic exploits of veteran North British 0-6-0 No 673 Maude of 1891 (BR No 65243), which gallantly steamed from the Scottish Railway Preservation Society Falkirk depot to Rainhill via the Glasgow & South Western and Settle & Carlisle routes towing two Caledonian Railway coaches. During its marathon journey home Maude poses alongside Class 85 electric locomotive No 85038 at Carlisle on 31 May 1980.

POSTSCRIPT

The preceding chapters have followed closely the remarkable renaissance of the steam locomotive in recent times, after its fall from favour. Moreover the history of railways has turned full circle for today, as in the mid-nineteenth century, a proliferation of private railways have spread themselves across Britain with steam the predominant means of power. What other of man's creations can so liken itself to the proverbial phoenix?

But to the dismay of some, as on British Rail, the diesels have filtered onto the preserved lines. However, steam die-hards may breathe again, for in no way, as happened with big brother, is it the start of a take-over bid, but rather to complement the steam stud and, indeed as preserved items of machinery in their own right.

Long ago many preservationists realised the advantages of having a diesel shunter or two about the place to be able to manoeuvre stock at a moment's notice and for hauling works trains. At the same time some, like the Keighley & Worth Valley and West Somerset Railways, found that a diesel multiple-unit or railbus could be run economically during off-peak periods, both for visiting tourists and the convenience of local passengers alike. This type of train has also proved ideal for viewing some of England's finest scenery, which abounds these lines. Dart Valley Railway also hopes to provide more early and late season trains with a dmu on its Kingswear line at times when steam would simply not be economic.

Students of BR will know that already large numbers of the diesels that displaced steam in the late 1950s and 1960s have themselves been superseded, some after a life-span of only 10 years or so – for longevity hardly to be compared with the majority of steam engines pictured earlier. Classes such as the Hymeks, Warships, Westerns, Deltics and Claytons, as far as BR is concerned, are part of history, but in this domain too the preservationists have been hard at work and a creditable number have been reprieved from the cutter's torch.

Needing a home for their acquisitions, the modern traction enthusiasts turned to the established private railways and, in some instances, received a sympathetic ear. Now they are valued at their new dwellings not only as useful stand-by locomotives, but also as a way of enticing the younger railway enthusiasts for whom providence decreed that they were born too late to remember steam in everyday service on BR. On some lines, in particular the North Yorkshire Moors Railway, the diesels find regular employment, while recently special diesel day galas have become annual features at a number of locations.

It is good that examples of modern BR traction should stand beside their steam counterparts (the National Railway Museum being well to the fore in this development), the lessons of earlier years, when many species of steam locomotives were lost beyond recall, having been well and truly learned. But no matter how the internal combustion engines are looked upon, undoubtedly the descendants of Trevithick's invention will remain the prime attractions, for what finer sight is there than a large steam locomotive straining at the bit, its exhaust ringing in the ears as its metal coupling rods force the wheels forward against a steep gradient. Nothing can compare or ever take its place.

The road ahead, as those who work on the private railways (either in a voluntary or paid capacity) will testify, will not be easy but, dare one say it, that the iron horse, unlike the dinosaur, will never become extinct. Surely man has become too fond of it for that to ever happen.

The words of John Keats sum everything up admirably:

'A thing of beauty is a joy for ever:
Its loveliness increases; it will never
Pass into nothingness.'

Need more be said . . .

127

Lion *in Silhouette.*